Big Research Questions about the Human Condition

Big Research Questions about the Human Condition

A Historian's Will

Arne Jarrick

ANTHEM PRESS

Anthem Press
An imprint of Wimbledon Publishing Company
www.anthempress.com

This edition first published in UK and USA 2022
by ANTHEM PRESS
75–76 Blackfriars Road, London SE1 8HA, UK
or PO Box 9779, London SW19 7ZG, UK
and
244 Madison Ave #116, New York, NY 10016, USA

First published in the UK and USA by Anthem Press in 2021

Copyright © Arne Jarrick 2022

The author asserts the moral right to be identified as the author of this work.

All rights reserved. Without limiting the rights under copyright reserved above,
no part of this publication may be reproduced, stored or introduced into
a retrieval system, or transmitted, in any form or by any means
(electronic, mechanical, photocopying, recording or otherwise),
without the prior written permission of both the copyright
owner and the above publisher of this book.

British Library Cataloguing-in-Publication Data
A catalogue record for this book is available from the British Library.

Library of Congress Control Number: 2020949958

ISBN-13: 978-1-83998-530-0 (Pbk)
ISBN-10: 1-83998-530-5 (Pbk)

This title is also available as an e-book.

CONTENTS

List of Illustrations   vii

Acknowledgements   ix

I. QUESTIONS AND ANSWERS – BACKGROUND, MOTIVATIONS AND AIMS   1

II. SUGGESTED QUESTIONS   15

1. What Explains That Some Kinds of Knowledge Are Widely Accepted Whereas Other Kinds of Knowledge Are Rejected?   17

    The question   17
    A knowledge society – what is one and are we in one?   18
    The decisive criterion: A knowledge-affirming attitude   22
    The state of the art and suggested steps forward   25

2. Why Do Some Societal Processes and Phenomena Develop in a Circular or Repetitive Way Whereas Other Processes Evolve along a Cumulative Trajectory?   29

    Cultural evolution   29
    Cumulativeness   31
    Non-cumulativeness   34
    What explains the difference?   35

3. Why Do Social Norms Change, Despite the Fact That Their Mission Is to Be Sustained? What Role Do Non-Conformist Individuals and Minority Groups Play in Cultural, Cognitive and Normative Change?   43

    A: Why do social norms change, despite the fact that their mission is to be sustained?   43
    B: What role do non-conformist individuals and minority groups play for cultural, cognitive and normative change?   46

The question – an introduction	46
State of art	47
The significance of outsiders for cultural evolution	52
A possible design	52

4. Does a Gradual Extension of Our Lifespan (and the Rise of Welfare) Imply a Growing or Declining Ability to Postpone the Satisfaction of Our Needs and Desires?   55

The relevance and importance of the question	55
The state of the art	59
What can be done?	64

5. What Explains the Widespread Diffusion of Inequality and the Gradual Emergence of Egalitarianism Over the Centuries?   67

Introduction	67
The equality–inequality gradient	68
The trajectory of egalitarianism	74

6. Why Do People Appropriate Aesthetic Experience (Both as Producers and Consumers of Cultural Manifestations), and What Are the Individual and Societal Functions of Such Experiences?   79

The issue	79
The essential questions	83
The state of the art	85
Ideas	87

III. WHAT LIES AHEAD?   89

Thematic clusters	89
The omitted questions	90
What the humanities are and what they are not	91

APPENDICES   99

A: *Five Thematic Clusters Summarising a Workshop on Big Questions*	101
B: *Translated Highlights from an Article on the Big Research Questions*	103
Notes	113
References	131
Index	143

ILLUSTRATIONS

Figures

1. Complexity as a function of coordination and differentiation   30
2. A Lorenz diagram   69
3. Inequality trends in Europe in the long run   73

Pictures

1. Intellectual innovations   32
2. Bicycle technology from the eighteenth to the twenty-first century   38
3. Striptease, by Marie-Louise Ekman   81
4. Shoes with red laces   82

ACKNOWLEDGEMENTS

This study was done at the Centre for the Study of Cultural Evolution, Stockholm University, founded in 2007 by ethologist Magnus Enquist and me. It is a truly interdisciplinary milieu, hosting researchers from very different disciplines and faculties, such as biology, mathematics, archaeology, linguistics, history and so on. My long-term interaction with people at the Centre has had a great impact on my intellectual orientation and has been essential for the progress of my treatise. I am especially thankful to Magnus, first of all for the profoundly thought-provoking discussions we have had ever since 2000 when we started a project on theories of culture, funded by the Swedish Riksbankens Jubileumsfond (RJ), and second for the extra money, granted by the Wallenberg foundation, he has provided for the final funding of this publication.

Also other colleagues and friends have contributed with substantially useful comments and suggestions on numerous aspects of the project. My most important interlocutor has been historian Janken Myrdal, my friend and colleague since the early 1970s. I have learnt immensely from my never-ending conversations with him, an impressively learned as well as a uniquely ingenious mind – and I am still learning from his reflections, often coming from completely unexpected angles.

The project started as a joint initiative between us, aimed at overcoming the fragmentation characterising the humanities today – in Sweden as elsewhere. As mentioned in Chapter 1, the first thing we did was to organise a workshop where each of the about 15 humanities scholars were invited to suggest 2–5 profound but insufficiently addressed research questions about the human condition. The workshop, generously funded by RJ and The Royal Swedish Academy of Letters, History and Antiquities, was very fruitful. It resulted in 50–60 suggestions from the participants. I am very grateful for the collection of innovative suggestions they offered – it has been crucial for the continuation of the project. From this rich intellectual repertoire Myrdal and I condensed 15 overarching questions which we presented in the Swedish academic journal *Respons*. This would not have been possible without the thorough and adequate minutes taken by the classical archaeologist Lena Johansson de Chateau, as well as by archaeologist Kerstin Lidén.

Business economist Kerstin Sahlin and historian Poul Holm have read the entire manuscript. They have given very different but equally useful comments and suggestions. Writer Per Molander, with a background in mathematics, shared his distinct thoughts on my presentation of his and other researchers' analyses and discussions on equality and egalitarianism. Theatre historian Karin Helander delivered friendly but clear-cut critical views on my discussion of aesthetics in Chapter 6. I am also grateful to linguist Marianne Gullberg for very inspiring feedback to the section on the humanities in the last chapter, and to psychologist Torun Lindholm for very good comments on an early version of the second part of Chapter 3.

Some years ago Myrdal and I presented our thoughts on the big research issues to historians at Lund University. Their sceptical but thoughtful reactions have been built into the present study. The same applies to the mostly positive reactions from historians discussing our presentation on a seminar at Åbo University. I have also presented our mission at various international occasions: the global humanities conference, Hanover, 2014; a seminar at the University of Campinas, Sao Paolo, Brazil, 2016; the UNESCO conference on humanities, Liège, 2017; and an international SIDA (Swedish International Development Authority) conference on the humanities, Stockholm, 2018. On the whole I have received reassuringly positive feedback from all these events, but also some criticism that has forced me to reconsider some of my thoughts. I am grateful to all these critical interventions, which I hope contributed to improve the study, whose remaining shortcomings I am of course exclusively accountable for.

I also want to thank Dag Retsö, Department of Economic History, Stockholm University, for the permission to republish free of charge 'The Pressure to Conform, the Need to Rebel: A Historical Project on Resisting Group Pressure' (see reference list) as one part of Chapter 3. I am also grateful to the GUNi network in Barcelona for their permission to reuse for free a few paragraphs of 'Knowledge Resistance: A Global Challenge – in Research and Education, in the Humanities and Elsewhere' (see reference list) appearing in Chapter 1.

At a late stage The Royal Swedish Academy of Letters, History and Antiquities generously decided to grant money for an open-access edition of my book, which I think is key to the spread of my message.

Last, but absolutely not least, I want to express my warmest gratitude to Anna-Lena Löfberg for indispensable and very generous help with all sorts of practical and legal issues, such as copy rights et cetera, but also for very good advice on numerous pressing matters at the very last minutes of this project.

I. QUESTIONS AND ANSWERS – BACKGROUND, MOTIVATIONS AND AIMS

I have long pleaded that humanities scholars should intensify and improve their efforts to try and find distinct answers to their research questions. They need to sincerely crave real results, that is to say, to advance knowledge. If they are successful in their endeavours, it means that they manage to learn something about the phenomenal world that they did not know in advance. I think that humanists should regard such new pieces of knowledge as *findings*, since this is just what they are. Well, *discoveries* would be just as appropriate. There is nothing dramatic in a claim like this. A scientific finding (or discovery) is simply something specific that has been *found out* about the world – be it inside or outside the living creatures inhabiting that world.

Once they gain real results, humanities scholars should also present them in terms of results (that is to say with claims to truth) and resist a certain fashionable temptation to degrade them into personal interpretations or biased perspectives that are dependent on their subjective vantage point or something similar that tends to belittle – although in their view ennoble – what they have accomplished.

Why do I emphasise the need for such a seemingly self-evident attitude to science and research in the humanities? Simply because it is not as self-evident among humanities scholars as one would wish. Far too many of them are unaccustomed to thinking, writing and talking about their achievements in terms of distinct scientific findings or results. When asked to mention important knowledge gains in their field of research, disappointingly many have nothing (or next to nothing) to offer. The interview quoted below is a typical example. It is a translated and slightly edited extract from a recent investigation into the condition of the humanities at a regional university in Sweden:[1]

> INTERVIEWER: Do you know of something from the humanities field of research that would count as a counterpart to the discovery of the Higgs particle in physics?

INTERVIEWEE: That requires more thought. Such a discovery probably gets lost in the great noise.
INTERVIEWER: Do you mean that it gets lost in the noise or that there just are no such findings?
INTERVIEWEE: I think there are such findings, but I believe that they are difficult to come across.
INTERVIEWER: There is a study claiming that, ceteris paribus, elderly people who walk fast live longer than people who walk slowly. It is a small-scale finding, but it is distinct. Could you mention something similar from the humanities?
INTERVIEWEE: I have no immediate answer. It is difficult.

Every now and then, the pretended crisis of the humanities is debated in the media. In a recent debate on this tiresome issue (on this occasion it concerned a humanities scholar's unfortunate absence in public debates), the disagreeing participants both took it for granted that humanists do not and should not deal in findings (*rön* in Swedish).[2] Indeed, an embarrassing consensus!

Furthermore, many researchers in the humanities seem not only unused to questions about findings, they even explicitly deny that their essential task is to be truth-seekers. The very notion of truth makes them feel uneasy and triggers many of them to make circumventing manoeuvres to avoid being ascribed such aspirations.

This is indicated by a series of more than ninety interviews with senior humanities scholars around the world, presented in a little volume called *The Humanities World Report* (2015).[3] Among other things, the interviewees were asked to give some examples of important findings gained in the humanities. By asking this, the authors wanted them to report on knowledge gained in the humanities in general, as well as in their specific field of expertise. The purpose was simply to get a good and useful collection of examples of what we know today about the 'human condition' thanks to research in the humanities that we did not know yesterday. Disappointingly, a substantial minority rejected the relevance of the notion of findings for the humanities, either by raising explicit concerns about the very notion itself or by trying to bypass the question altogether.

For instance, one of the respondents discarded the question by stating that 'the humanities should [not] talk in terms of "findings"'. Similarly, another respondent said, that he 'would want to point to insights gained and ways in which society and culture have been enriched by the humanities rather than point to "findings" per se'. In the following example of an answer to the question, the same reluctance to provide specific examples is expressed: 'But in many humanities disciplines, what you do is to "think around" a subject (e.g. a historian is not trying to answer a question like: "What were the five causes of WW1?").'

Likewise, while one of the respondents mentioned a couple of very down-to-earth examples, such as the fact that 'the intrusive legislation of the 1530s may have had a direct effect on the development and adoption of certain literary forms in the period', only to add, in a typically humanist fashion, that 'these "findings" [again these self-destructive scare quotes – AJ] are likely to be contested by the next scholar to examine those texts, almost as a matter of course'.

Another way of circumventing reporting on findings, without discarding the matter as such, is to redefine findings in the humanities as *finds*, 'e.g. making a text available through translation, deciphering scripts', as said by one of the interviewees. In my view this is something else, although as important as findings. This is to conflate unearthed traces of the past with empirical, more or less general statements about it based on these traces. In my view, a finding is not only a matter of fact as such, but a statement about this matter. The most pregnant refusal to accept the notion of findings was the following one:

> I do think that this runs counter to our sense of the humanities as a dynamic discipline, and we should refuse to answer such queries because it puts the humanities in competition with, and defensive about, the knowledge that is generated by the natural sciences. Of course, we know much that we did not know before because of humanities research, but the most important lesson we have from the humanities is that we can still keep thinking about what we know, and see if we can un-know it, un-ravel it in some way, or build upon it.

It must be added, though, that when the initial interview question was rephrased, the response shifted considerably among the remaining interviewees. When *finding(s)* was replaced with *knowledge progress*, a large dissenting minority turned into a large assenting majority. *Knowledge progress* was acceptable, *findings* was not – odd as it may seem.[4]

Now, I still do not know whether such a positive response to the rephrased question also implies a truth-affirming attitude. In any case, if such an attitude were generally applied, it would improve not only humanities research but also its reputation among other scholars, as well as among the public. It would be a good thing. It would, though, not be good enough. Why? Because for the answers to be distinct, the questions must be distinct too, and today this is generally not the case to a sufficiently high degree.

My basic message can be put in a straightforward way: questions need to be as clear and simple as possible in order to enable unambiguous answers.

Simple without being simplistic, nuanced without being embroiled – that is the ideal. Unambiguous answers (not to be confused with irrefutable answers) are much wanted, although not always possible to attain. Moreover, if one wants the questions to be highly significant for the understanding of the human condition, there should not be too many questions. Even in this respect, there is much to be wanted in today's humanities research. Instead of gathering around a limited set of profound questions and holding on to them until the answers begin to appear, as a whole, the humanist guild scatters its scientific energy on too many disparate things – replacing them far too often with hundreds of new questions, 'perspectives' and 'problematisations'. In its turn, such a research culture may hamper a cumulative growth of knowledge, the possibility of which, moreover, is regrettably often denied or even viewed with suspicion.[5]

In the past, humanities scholars did not shy away from drawing the big pictures. Admittedly, they were not very often more than half-heartedly interested in the empirical basis for their great thoughts. Nevertheless, their concentration on the fundamental aspects of human life was worthy of following. Unfortunately, however, some thirty years ago the postmodernists appeared on the scene, attacking the grand old syntheses by reducing them to fairy tales, to 'narratives' or 'grand narratives'.[6] The postmodern sceptics were certainly not themselves substantially empirical, sometimes they were even rather anti-empirical and based their iconoclasm against the modernist masters on the denial of the possibility to analyse the processes of human life in general terms. By doing that, they succeeded in refunnelling many researchers' efforts away from such an ambition towards the bricolage of topics that they already claimed to be characteristic of the humanities. On the other hand, however, they forced many researchers to become more meticulously empirical, which was a good thing, although at the same time they brought them too much down to earth.

Against this background, the task of this book is to pave the way for a return to the good old modernist ambition to formulate and answer the really big research questions about the human condition, without dispensing of the high empirical demands that should be met by such an enterprise to any degree. Thus, the thematic fragmentation characterising the humanities, partly conditioned by the postmodern de(con)struction of synthesising ambitions, is one of my overarching motivations for raising this task. It should be stressed from the outset that I will solely raise this task by suggesting questions and possible designs with which to address them, not by providing answers to them. Answering them is up to anyone enticed to respond to my quest.

To ask big questions about the human condition implies a quest for generalisations. Is this not inappropriate in a field held to be characterised by extraordinary variation? No, it is not. The overarching purpose of all science is to try to discern patterns in phenomena of whatever kind, whether human or non-human. What does that mean? It means to look for structure in a seemingly unstructured world, to identify what recurs in that which seems to vary limitlessly. It means to go beyond the registration of the specific in an attempt to formulate general truths.

Doubtlessly, to discern patterns – that is to say, to identify what is common at the expense of differences – is to simplify what appears complicated. Is it really acceptable to do so? Well, it is not only acceptable, it should even be endorsed. Although this can never be everybody's task, it is the major mission of the scientific system as a whole. It is certainly not the only task, and it does not imply any denial of the ever-present variation in the phenomenal world. In the last chapter I will discuss why generalisations are compatible with the recognition of variation (95–96).

What I suggest is certainly not an easy task to carry out. The more wide-ranging the questions, the trickier it may be to make them distinct and the more difficult they are to answer. Hence, my quest for a return to both big questions and answerable questions may seem untenable, even self-defeating. But although difficult, it is not an impossible mission, and since it is badly needed, it is worth trying. Furthermore, once we know what we want to know, we should also take pains to operationalise our overarching knowledge interest into concrete and researchable questions. Surrendering to narrowness in the face of roadblocks on the way from the general to the concrete is thus no alternative.

Now, in real life we do not always start a scientific endeavour with perfectly crystal-clear questions in mind, but rather often we start with a vague knowledge interest that is in need of specification. This is just as normal as the opposite situation and it is not a big problem provided that we manage to make our questions distinct at some point in the process. The specification of the question at stake is often an iterative process no less than the systematic attempts to answer it. As often as not, the two processes are intertwined.

But sometimes we do really already know what we want to know from the very beginning. In such unusually lucky circumstances, we may manage to set up simple and seemingly precise questions. However, what seems strikingly clear at the inception of the research process often turns out to become gradually less clear along the road. This happens because of unexpectedly confusing or annoying experiences in the archive, the field or the lab, forcing us to recognise that the matter was not as simple as it appeared a priori. Generally speaking, it happens because research is basically a journey into the unknown, and thus destined to bring us to unwelcome surprises, as well as welcome surprises. Such progressing confusion is nothing to worry about once it triggers us to re-clarify our aims and questions. But it in no way allows us to dispense with trying hard to both ask and be guided by as simple and clear questions as possible. This remains the goal whatever, a goal far from accomplished. Let me expand a little on this troublesome state of the art.

As illustrated above, whether shunning the matter or not, it seems that humanities researchers are badly trained or unused to talking in terms of substantive results and in terms of findings.[7] This is a problem for more than one reason. First, if humanities scholars do not have a clear intention to present their research in such terms, there is probably a risk that they do not even organise their research in order to ask clear (= answerable) questions and to arrive at distinct answers to them. Second, by shying away from presenting the outcome of their research as clear results, they also miss the opportunity to be properly scrutinised by their peers. All scientific results are fallible, but only clearly stated results can be falsified or validly resist falsification. Instead of being properly tested, vaguely presented research lends itself to never-ending disputes, to opinionated 'scientific' fights leading nowhere. Third, to publicly proclaim that we do not produce as reliable knowledge as other scientists is detrimental to the citizen's trust in the humanities. It conceals the fact that we do produce real knowledge.[8]

This could be put in a very simple way. Since all research starts with an interest to get to know something that you do not already know, its point of departure is a question – or a set of questions. To me it seems natural that once one has asked a question, one also wants to get an answer to it, although it is also nice to generate a new one, basically in order to contribute to the never-ending production of knowledge, but, to be honest, sometimes just to sustain one's research activity (and earn a living). Why else ask it?

Is this needless to point out? Well, not in view of an often-expressed attitude that it is more important to ask questions than to answer them. Furthermore, some humanities scholars think that a question that could be answered with a *yes* or *no* is badly put. I do not understand why.

Given that you have a quest for answers to your questions, normally you will regard some answers as better than others. This means that you believe the better ones to be true in some sense of the word. Why else consider them better? This being the case, you should also be ready to *claim* them to be true, although at the same time humbly admitting that they may turn out to be less than completely true in the long run. However, if you have tried diligently but in vain to falsify your results, they may resist future attempts at refutation. As I see it, such statements are statements of findings, whatever they are about. It is as simple as that.

As long as many humanities scholars oppose the above reasoning, there is a mission. This implies a quest for a unified culture of science. It purports to overcome the gap between the two cultures that C. P. Snow once compared, discussed and to some extent complained about. Let alone that he asymmetrically contrasted *research* in the sciences with *cultural activities* among humanists, not with corresponding scientific research in the humanities.[9]

There are other motivations too for my mission. One such motivation is the disturbing fact that policymakers more and more often impose short-sighted research commissions on academia, in the humanities no less than in other sciences.

Hopefully, by bringing them to believe both that humanities scholars are willing to identify the most important questions and that we possess the scientific tools to carry out this task, we might be able to calm them down a little and encourage them to refrain from such meddling into our business. I am hopeful, since I believe that politicians are able, without too much effort, to recognise that in the long run, the most scientifically profound questions are also the most societally relevant ones.[10] The task then is just to go ahead and formulate some questions of this sort, to indicate where we should be heading. This is precisely what I will do in this book. Of course, my suggestions are neither complete nor unquestionable. They should be regarded as examples of possible ways forward or possible types of way forward. Since I am far from perfectly well informed about the state of art in all fields to be surveyed and discussed below, I welcome any expert in any of those fields to correct me if and where I am mistaken about the proper way forward.

One of my sources of inspiration has been formed of the questions once raised by mathematician David Hilbert. In the year 1900, he presented 23 at that time unsolved mathematical questions to which he wanted his colleagues around the world to provide correct answers. Shortly after, he presented 10 of them at an international conference in Paris, most of which mathematicians have since managed to solve. I appreciate Hilbert's initiative, which in due course has turned out to be a great success. I would like to contribute to something corresponding to that, namely to encourage researchers to make concerted and focused efforts aiming at a deeper understanding of the human condition. I believe that such efforts might bring the humanities considerably forward, although I am fully aware of the fact that the answers to the questions I will suggest below can never be as conclusive, unambiguous and incontestable as the answers to Hilbert's mathematical questions. Actually, this sets mathematics apart from most sciences, not only the humanities.

As a precaution against a possible misunderstanding of my mission, it needs to be underlined that the intention is not to urge all humanities scholars to confine themselves exclusively to the topics addressed and suggested in the discussion that follows. Such a totalising attitude would be devastating for the intellectual destiny of science and research, not to mention the futility of imagining that a whole scholarly community ever would comply with such a request. Indeed, a nightmare – and as unreal! In the long run, academia and society at large have a lot to gain from sole riders preoccupied with questions that most of their contemporaries – be they researchers or not – do not understand the meaning of or consider odd, unimportant, incredibly esoteric, outright mad and what not. Ideas that at a certain point in time appear as totally flawed may, in the long run, turn out to be the germ of a scientific breakthrough. In the even longer run, they may serve as keys to the solution of many societal problems and lead to the fulfilment of many basic human needs. In addition, it is perhaps needless to point out that my suggestions are just a few of many other possible questions. Of course, they cover far from all the questions that should be addressed.

Since a short-term perspective means that it is not always easy to distinguish crap from gold, the useless from the useful, sterile from ingenious outsiders, academia has to provide substantial space for seemingly odd people to devote substantial time to odd issues whose significance may appear unintelligible to those in charge of the resources. Of course, at the end of the day, a lot of these odd ideas will be abandoned and for good reasons. But the same will be the fate for many mainstream ideas. Waste can nowhere be completely avoided – it is even a necessary ingredient in all creative activities. Failure breeds success.

In other words, a balance needs to be struck between concentrating on a limited number of essential questions and safeguarding a haven for idiosyncratic research interests. Such a balance should be established bottom-up, that is, by the research community itself safeguarding that it will be done on scientific grounds. This being said, I still think that in the present situation, thematic fragmentation is the main shortcoming to come to terms with.

Before proceeding to the criteria I have tried to apply in the following chapters, I would like to share my speculative thoughts on why so many humanities scholars display such a reluctance to ask clear questions, why they avoid presenting their results in terms of results and why there is such theoretical, as well as thematic, fragmentation in the field.

In physics everyone has to relate to the grand theories of relativity, electromagnetism and quantum mechanics, either by applying or by developing (and sometimes, but rarely, revolutionising) them.[11] The same applies to evolutionary biology, where no one ignoring the Darwinian-Mendelian synthesis will be taken seriously.[12] Genetics, at least the theory of heredity, is another case in point, recently 'supplemented' by epigenetics. From this it follows that most researchers in these fields are preoccupied with the same basic issues, although the specific empirical questions addressed may and do vary a lot.

In contrast to this, in the humanities there is not much in the way of a profound master theory that guides the intellectual activities of most researchers.

This is certainly so in many scientific fields apart from the cases just mentioned. But this state of affairs is probably more pronounced in the humanities than elsewhere, making it difficult to draw a general picture of what goes on there.

Yet, the lack of a sustained common theoretical orientation among humanist researchers does not prevent them from being attracted by and putting forward certain generalisations about the human condition – which they should not and cannot avoid anyway. But again, these theories are rarely as long-lasting as the timelessness their generalisations might imply. Perhaps, it is even the very lack of a central theory in the humanities that explains that the theoretical turnover is quicker here than in some of the sciences. Despite having a great impact on most researchers, such theories tend to come and go – and come back again. They are quite often described as certain 'turns'.[13] A very typical example of this is the field of aesthetics, specifically addressed in Chapter 4. Although such theories may attract many humanist researchers for some time, they rarely enforce themselves on each and every one of them as, for instance, quantum mechanics does to physicists. More so, the less they are anchored in solid empirical ground.

Let me just add that I do not claim that theoretical stability is always better than theoretical flexibility. There needs to be a balance between continuity and renewal. Sometimes scientific progress might be stimulated as much in a joyful setting where researchers are invited to play freely with different theoretical options as in research environments where basic theories have to be taken as givens.

Now, be this as it may, due to what has been said so far, knowledge production in the humanities does not appear to be as cumulative as in other sciences – it appears less so than in fact is the case. Many humanists seem to conceive of the humanities rather as a non-cumulative enterprise, claiming that each new generation of scholars in some sense starts (and has to start) afresh instead of building on what has been accomplished by the previous generation. What has been erected by one generation of researchers is expected to be erased by the next and so on – so goes the gloomy 'narrative' about the predicted fate of one's achievements.

But this story is less true than humanists themselves think. It is largely a narrative in the bad sense, a fairy tale, far from literally true. Luckily, a stock of knowledge is gradually being built here too. I will make a little detour to give some examples of reliable and new knowledge gained from very different areas of research in the humanities.

Some twenty years ago a research programme was launched on Sweden's relations to the Nazi regime during the Second World War. One of its major results can be succinctly summarised as follows: the Swedish government had the intention to remain neutral to all parties involved in order to remain outside the war and thus preserve the continuation of Sweden's extraordinarily long peace period (from the end of the Napoleonic war in 1814). Unfortunately, these two intentions turned out to be incompatible. While Sweden succeeded in preserving its peace, it gradually became less and less neutral, feeling the pressure to submit to German demands in a lot of dubious cases. Peace was bought at the expense of neutrality.[14]

This is an example of solid empirical results without far-reaching generalisations. This is good enough, resembling what goes on elsewhere. But of course, like researchers in other domains, humanities scholars also have the ambition to go beyond the mere discerning and reporting of facts in order to reach different levels of generalisation.

It has been shown beyond doubt that homicide rates in Europe have declined substantially over the centuries.[15] Linguists have made experiments strongly suggesting that it is as easy for an adult as for a child to break into a second language.[16] By carrying out longitudinal studies, psychologists have shown that elderly people's both semantic ($\approx$ knowledge) memory and episodic ($\approx$ direct experience) memory are far better preserved than has been previously claimed on the basis of cross-sectional studies.[17] Recent studies of cognitive development in the beginning of life have brought very interesting results, enforcing a substantial upgrading of our view of the cognitive capacity of infants. For instance, it has been shown in experiments that an infant of six months of age already moves its eyes to the mouth of another person when that person starts picking up a spoon from the table. Thus, by anticipating what will happen next, the little child reveals an ability to decode another individual's intention.[18] Perhaps this could be linked to the notion of joint attention, the species-specific ability of two humans to jointly focus on something else than each other, some 'third' thing.[19] It has been demonstrated that the passage from authoritarian rule to democratic rule is normally a rapid process. This adds to two other stable and general observations provided by research on democracy: first, that the movement towards democracy is most often violent and, second, that, once established, democracies do not wage war against each other.[20]

More results on this level of generalisation will be hinted at in my discussion below, but these examples already represent robust and reliable results of serious scientific efforts. Thus, as elsewhere in academia, there is a lot of established and no longer questioned knowledge that humanist researchers cannot avoid drawing on. However, due to the largely unfounded self-belittling attitude, the cumulative progress of knowledge is less systematically developed in the humanities than in other fields of research. In that respect, the rejection of cumulative progress shown by some trendsetting scholars is somewhat self-fulfilling. It also probably contributes to the remarkable reluctance among humanities scholars to talk about the results of their scientific efforts in terms of findings. Once you acknowledge your achievements as findings (i.e. as reasonably reliable new knowledge), it is reasonable to expect others in your research field to build on them – whether critically or not. On the contrary, if you devalue the fruits of your intellectual efforts by just calling them 'perspectives' or even just a kind of 'personal perspectives', you cannot count on being taken seriously or the knowledge being regarded as knowledge.

Let me now present my criteria. In order to formulate as productive and distinct questions as possible, I think that the questions should
    be of paramount importance for the human condition
    be unanswered to a high degree
    be possible to answer
    enable different answers.

The first criterion is in no need of justification. What else could big research questions concern than precisely big questions, those that either haunt or bless humankind?

The second criterion may also seem self-evident and thus needless to even mention. But it is not. Over the years I have gathered massive experience of evaluating research proposals where too many applicants have declared their final theoretical position in advance – that is they have declared the perspective they have decided to apply in their upcoming empirical study. Thus, what one would expect to be the outcome of the enterprise is rather its point of departure. Sometimes such a procedure is perfectly legitimate, for instance, in those parts of philosophy where the aim is to spend the allotted time precisely in order to prove a certain theorem. But normally it is not legitimate, and in many cases it tends to foreclose true curiosity instead of stimulating it. Further, my intention has never been to cover all of (or even most of) the essential questions about the human condition but only to cover those that have so far been either neglected, insufficiently addressed or unsatisfactorily answered. Of course, a lot of important human matters have already been exposed to scientific scrutiny and will therefore be omitted here. I will come back to them near the end of my discussion, while also briefly mentioning topics that would have deserved being addressed here though I have not been able to do so, mainly due to my lack of stock-taking of the ever-widening field of humanities.

One more important remark should be made regarding the second criterion. At the outset, when historian Janken Myrdal and I tried to identify what research on the human condition humanities scholars should focus on, we based our provisional recommendations on bold statements about highly neglected fields of research. However, as I have familiarised myself with the state of the art in one field after the other, quite a few of those fields have proven less virgin than we thought in the first place. In some of the cases we were right, but in others our complaints turned out to be exaggerated. Yet, since none of the suggested questions have received sufficient answers so far, there is still a big need to address them and to do so with more concerted efforts than have been made to date.

Regarding the third criterion, it should be pointed out that some questions evoke people's intellectual curiosity to such an extent that they cannot resist the temptation to start working on them, regardless of whether or not the questions in fact are researchable. They just get carried away. To follow such an inclination is often a good thing since it may be impossible to know beforehand whether the question opens a new avenue of research or leads to a dead end. However, at some point one will have to decide either to continue or to back out, and at this point the scientific feasibility of the question needs to be assessed. In this endeavour, I have tried to establish this in advance. That is the rationale behind the third criterion.

Fourth, the questions must be open-ended (i.e. genuinely allowing different answers). If not, they are just pseudo-open questions and, as such, they would provide a way to get around the second criterion listed above.

These are the criteria I have tried to apply in the following chapters. In the search process, we operated with around fifty potential questions. They emanate from a workshop in 2015, organised by Janken Myrdal and me, to which around fifteen scholars, mostly from the humanities, were invited to suggest two to five topics each. As expected, quite a few of the questions were overlapping. After the workshop we excluded some, merged some and first reduced them to a manageable number of fifteen questions and eventually down to nine, of which finally only six will be treated in detail. To some extent, but far from complete, they resemble the five clusters of suggestions I already tried to discern on the second day of the workshop (see appendix A).

A few years ago, the 15 questions were presented in an article published in a Swedish journal.[21] Later on we made our thoughts available in English, the highlights of which are to be found in appendix B of this book (103–12). In this article, the questions were deliberately ordered in an arbitrary way. There were two reasons for this procedure. First, we wanted to avoid the impression of any hierarchy between the questions. Second, we tried to formulate the questions so that each of them could stand by itself and be addressed independently of the other questions, irrespective of the fact that there are quite some overlaps between them. However, considerably fewer questions will be discussed here, and it turns out that each of them can be seen as one among a number of potential instances of a type of question. Therefore, I have decided to arrange them here in a more logical order, mainly due to their level of abstraction. Thus, I start out by addressing the question about knowledge resistance, which could be regarded as a meta question from a societal perspective, and then become more narrow and specific along the road. Why a meta question? Because to advance knowledge is largely meaningless in a world of ignorance or outright disrespect for knowledge. Normally, it is no problem that people in the short run often display resistance when they come across new knowledge that is counter-intuitive in relation to their deep-seated cognitive habits. This is just normal along the trajectory of knowledge progress. The big problem is sustained ignorance, people's perseverance in turning their back on real and important knowledge gains. What would be the meaning of our scientific efforts in such a stubborn anti-intellectual climate? Wouldn't it alienate even the most curious of minds?[22]

That the questions differ much in degree of abstraction resembles what was also the case with Hilbert's questions. Moreover, the varying length of the chapters does not signify any difference in the importance of the questions. All this may make the six remaining questions appear arbitrarily chosen, first in the sense that they relate to the arbitrariness of my competence and, second, because they differ in regard to the level of aggregation or abstraction, but also in regard to how much I go into detail about a possible design of future research. And so they are.

However, the perspective applied is not arbitrary; it is the historian's take on the big questions. It is even the approach of the world historian. This is so, first, since I think that almost all the questions would gain from being addressed from a long-term perspective, as well as from a worldwide perspective. Second, it implies that I recommend that researchers apply a comparative approach in their endeavours, both in the horizontal and vertical sense of the word (i.e. cross-culturally as well as cross-temporally). At the same time, many questions outlined in what follows would gain from being addressed with the concerted efforts of historians and other social scientists while systematic experiments, staged by psychologists and others, could help answer the questions.

Finally, it should be stressed that I strongly adhere to the unity of sciences, despite the fact that I have a specific point of departure, that of the world historian. This should, however, not be understood as the kind of scientist reductionism once implied in such a standpoint. As will be obvious in the discussion below, it is rather that I do not make any epistemological distinction between the human and natural sciences. This is absolutely not to deny that different fields of knowledge require different methods. I will come to this issue in the very last chapter of my treatise (the section headed 'What humanities are and are not').

II. SUGGESTED QUESTIONS

Chapter 1

WHAT EXPLAINS THAT SOME KINDS OF KNOWLEDGE ARE WIDELY ACCEPTED WHEREAS OTHER KINDS OF KNOWLEDGE ARE REJECTED?

The question

What is at stake in this chapter is my desire to understand and systematise the conditions for the spread and acceptance as well as the rejection of knowledge. The notion of knowledge is here taken in its broad and classical Platonian sense as justified true belief, still a workable definition. This implies a rejection of the relativist idea that conflicting truth holdings of the same phenomenon could coexist as different 'knowledges'. Witchcraft cannot have both existed and not existed. It was as false an idea in the seventeenth century, when most people believed that it was real, as it is in retrospect today – or will remain from any vantage point in the future.[1] Thus, knowledge is not just 'whatever is taken to be knowledge in a given milieu or culture'.[2]

The societal as well as the scientific significance of the question is obvious. Resistance to knowledge is as ever present in humankind as its restless quest for knowledge. To overcome such resistance is as pivotal for all science that intends to have an impact on society as it is for the destiny of humankind itself. Rejected or ignored knowledge, whatever its importance and quality, is of little use, and people making decisions on false grounds are potentially behaving contrary to their own interests and sometimes also contrary to the interests of humankind as a whole.

Is it possible to discern dissimilar or even contrasting intrinsic traits of knowledge that generally either invite its adoption or trigger its repulsion per se (i.e. irrespective of its specific cultural or historical context)? Self-evidently, counterintuitive knowledge is more difficult to assimilate than knowledge which suits people's preconceptions or ideological leanings. This is well known and may be covered by the psychological mechanism conventionally called 'confirmation bias'.[3] But could other such traits of knowledge be identified that affect its varying reception or impact? In addition, what is the significance in this respect of certain pivotal situations, such as certain societal atmospheres, certain human experiences or attitudes? In sum, what are the general cognitive, emotional and ideological factors that may help explain the adoption of knowledge as well as the repulsion of knowledge?

One obviously relevant circumstance is that the growth of the total stock of knowledge is far more rapid than the growth of what an individual human being has the capacity to incorporate. This means that the gap inexorably widens between the former and the latter. It might also speed up the incessantly ongoing process of specialisation, in its wake more than ever forcing people to select what knowledge to adopt and what to neglect or even reject. Moreover, there is a risk that it will reduce or block people's readiness to encompass an increasingly vast mass of information, or, even worse, alienate people from gathering knowledge altogether.

Below I will go into detail about the background to the questions at stake in this chapter.

A knowledge society – what is one and are we in one?

Over the last fifty or sixty years it has been repeatedly and frequently claimed that we live in a knowledge society, sometimes more narrowly labelled *a knowledge economy*.[4] In the 1960s, a growing number of social scientists, such as Robert Lane, Alain Tourraine and Peter Drucker, began to characterise contemporary society in this way.[5] Sociologist Daniel Bell is perhaps the most well-known exponent of this line of thought. In his seminal work on post-industrial society from 1972, he attempted to show that knowledge and the knowledge sector were growing exponentially in his time.[6] Although Bell's primary case was the United States, his ideas were rapidly and widely adopted as an accurate description of many other parts of the world.[7]

However, traces of similar ways of reasoning can be discerned among scholars far earlier than that, for instance in the writings of nineteenth century sociologist August Comte or as expressed by the mathematician and philosopher Alfred Whitehead in a little book published in the 1920s. But it was in the 1960s that this characterisation was made explicit. Gradually it became commonplace.[8]

In the modern twentieth-century discourse on the knowledge society, the 'we' who are supposed to live in such a society seem to mostly be citizens of the Western world, however, accompanied by culturally adjacent societies, whether geographically located near the Western world or not: Israel, Japan, Singapore and others. The fact that societies outside this domain are about to catch up is as feared as it is officially welcomed. Although societies of all times draw on knowledge of some kind, reliance on real and systematically advanced knowledge is claimed to be particularly characteristic of our time.

However, the meaning of *knowledge society* is as unclear as its pretended omnipresence has been distinctly and repeatedly claimed, disregarding the present fashion to replace it with the equally bold but equally poorly proven claim that we now live in a *post*-knowledge society.[9] How *knowledge society* should be understood has varied over the years and continues to vary between scholars. How has it been defined and how should it be defined?[10]

One defining trait that has been frequently applied is simply that the stock of knowledge is exceptionally large in our time (≈ from the mid-twentieth century) as well as expanding at a previously unforeseen pace. It is almost a matter of exponential growth, as stated by Bell and others. Knowledge is stored in various media outside the human brain, and the incessant introduction of new such media may indeed facilitate and speed up further knowledge expansion. But it is also claimed that due to the plasticity of the human brain, we have improved our ability to store and digest more and more knowledge within that brain. Among other things, this is indicated by the so-called Flynn effect, which signifies that IQ has improved considerably over the last seventy or eighty years.[11] And this has happened without any substantial change of the human genome. It is, thus, a matter of cultural evolution.

Measured by this simple, or even simplistic, definition, it appears indisputable that we do live in a knowledge society – or at least do so more than preceding generations. This is the case despite the fact that new findings are not only incessantly brought in but also continuously subtracted or lost from the overall stock of knowledge. It seems more than likely that the inflows far outdo the leakages and, furthermore, that some of the knowledge that has been abandoned or thrown into oblivion can be rescued from the darkness and reutilised, most often in new ways.

According to a different approach, it is the spread and distribution of knowledge, rather than its quantity or rate of growth, that are the decisive criteria. The more widespread and evenly distributed knowledge becomes the more society deserves being designated a knowledge society. This becomes even more the case to the extent that the authorities refrain from interfering with the streams of information by imposing this and banning that.

According to the second of the two definitions of *knowledge society* given above, it would be reasonable to conclude that ours is a knowledge society, although not entirely so. On the one hand, it would indeed be hard to deny that knowledge is today more widely and evenly distributed than ever before due to the explosive growth of mass education over the last century, as well as the rich repertoire of bottom-up initiatives taken by various popular movements and

NGOs over a corresponding number of years.[12] Moreover, the wide-ranging freedom of the press and other media has contributed substantially to an even wider dissemination of knowledge.

On the other hand, as the total stock of knowledge gets incessantly larger, its advancement has become more and more specialised, which, ceteris paribus, appears as an increasing obstacle to its digestion. More and more often in everyday life, the citizens of modern society have to rely on experts rather than on their first-hand or personally acquired knowledge. The experts themselves are no exception to this predicament – to be an expert means being a non-expert in most things. By providing shortcuts to knowledge, the experts enable us to utilise it without really understanding it. Moreover, as pointed out over and over, today's media – like the media of other ages – produces lots of misinformation, not only true knowledge. And despite obvious progress in the spread of knowledge, it is still unevenly distributed globally as well as between social classes and sexes.[13] So, in view of these simultaneously ongoing processes, what is the answer? Do we live in a knowledge society or in an expert society, or neither? Finally, there is presently a worrying global trend towards autocratisation in some formally democratic countries where the freedom of expression is put under increasing pressure and even about to get squeezed.[14]

A third definition is the stress on the necessity to know – and to know a lot – as an essential requirement for the citizens of modern society. Analytically, although not in real life, this criterion should be distinguished from the stress on mass of knowledge and from its diffusion. A high level of knowledge is considered not only an asset that enables people to get along in society but also a stepping stone to a good career, good health and a long life. Rising knowledge demands from the workforce, and the gradual decline of unqualified jobs are held to be the outcomes of the rapid technological development of industry and also of the growth of the service sector at the expense of a declining industrial sector. Here, the quality rather than quantity of knowledge is crucial, yet it is motivated more by business needs than by a quest for enlightenment and the democratic empowerment of the population.

Again, also according to the third definition, it could be held as true, yet again only conditionally true, that today we live in a knowledge society more than ever before. Certainly, as industry has become more and more technologised and the service sector more intellectualised, a growing majority of the workforce are expected to acquire matching, high-level skills. Even in Turkey, in this sense the least demanding country in Europe, no more than 15 per cent of the work force can dispense with the need to possess the high skills typical of modern life.[15] It has also been shown beyond reasonable doubt that the better people are at meeting this demand – that is, the demand for a higher level of education – the more they will be prosperous, healthy and long-lived.[16]

However, it is not clear to me whether these steadily increasing demands on the workforce are as demand driven as has been frequently and unanimously claimed to date. Could they not be output driven as well to a substantial degree? Is it really the complexity of working life that conditions these demands, or is it as much, or even more, the abundant supply of highly educated people that triggers employers to ask for them? If so, could it be the case that people in many occupations are overqualified – at least in this narrow sense? One indication of this is that politicians have recently begun to raise the need to offer simple jobs to badly educated people migrating to Europe in exceptionally large numbers – refugees and others – from countries not considered to have knowledge societies, facing them with hitherto unforeseen problems of integration. As if the pressure from the inflow of all these poor people leads to a rapid dissolution of the knowledge society, once considered so solidly established. A more robust finding pointing in this direction is the fact that the proportion of citizens with a low level of education has declined substantially more than the proportion of jobs requiring only a low level of qualifications – at least from the 1970s to the beginning of the present millennium.[17]

I do not know the answers to these questions. That is why they deserve being asked. However, I would not consider an answer to these particular questions decisive for the overarching question of whether we live in a knowledge society – whatever the answers might be. The reason is obvious: this is just one possible angle from which to approach the matter, and, as I see it, not even the most fruitful one.

According to quite another view, none of the three aforementioned definitions would qualify as sufficient to settle the issue. Basically, a society does not deserve the designation *knowledge* unless it is characterised by a widespread knowledge-affirming attitude among its citizens. According to such a view, it is the last-mentioned definition that should count as the defining trait of a knowledge society, whatever the amount or spread of knowledge. Rationally, knowledge-affirming people take steps to optimise rather than maximise their knowledge about the phenomenal world. They try to base their actions on true knowledge, not on wishful thinking. Ideally, they would behave as everyday Popperians in the sense that they would spend as much intellectual energy on critically examining their own beliefs as they do on examining other people's beliefs.

Although the four definitions discussed above are all interlinked, it is specifically the last one that brings me to the core of this chapter. It is closest to the question of what explains why some pieces of knowledge get appropriated or accepted whereas other pieces are met with repulsion or rejection. In this particular context I am primarily concerned with the attitude towards knowledge: how and through what mechanisms it takes shape and is sustained, modified and even refuted. This is regardless of what would be the most adequate general litmus test of a knowledge society.

The decisive criterion: A knowledge-affirming attitude

Do human societies looked upon this way match these criteria? From a macro-historical perspective and measured on the population level, the answer it is unequivocally *yes*. People today know substantially more about the phenomenal world than they did in the past, and they also possess a larger repertoire of cognitive tools with which to continue extending their sphere of knowledge in the future. For example, in the High Middle Ages it took about thirty to forty years to master the mathematics that today's high school students incorporate ten times faster.[18] Not to mention negative numbers, which not even the most eloquent mathematicians operated with in the sixteenth century but which contemporary schoolchildren almost unexceptionally just take for granted from the age of 10 (or earlier).[19] It is also likely that a knowledge-affirming attitude has moved forward alongside the steady progress of knowledge. This is the most important step forward against the backdrop of the specific discussion in this chapter.

However, by applying such a macro perspective, one may give the impression that humans of all ages have always had a straightforward and open-minded craving for all new knowledge, that they have unconditionally been ready to adopt it once they have managed to cognitively grasp the novelties of their times. This is of course basically false. Generally, humans do not only seek knowledge, they also seek to avoid it or just deny it. So, each stage of the overall long-term progress of knowledge and a knowledge-affirming attitude are interleaved with resistance to new knowledge but also to certain pieces of old knowledge. How come?

Normally, new ideas are born in the minds of peculiar individuals or as the offspring of the efforts of tiny minorities – discoveries no less than ideological ideas or innovations of fashion. No wonder then that new-born ideas are often met with suspicion or even outright hostility by their intended recipients, more so the more counterintuitive they appear, and even more so to the extent that they challenge people's most profoundly cherished and often culturally inherited beliefs. Nevertheless, solid discoveries tend to gradually break through the resistance, bringing people to eventually accept them – reluctantly and with certain delays. Moreover, in everyday life most people unhesitatingly make use of many things that are based on essential scientific findings of which they may be completely ignorant. The smart phone is an obvious example, a device that would not have been possible without Maxwell's discovery of electromagnetism. How many users are aware of that?

Altogether, this means that the majorities of today embrace much of what past minorities failed to get their contemporary majorities to adopt, such as heliocentrism (from Yajnavalkya to Galileo), seeing the bloodstream as a closed system (Harvey), electromagnetism (Maxwell), the fact that we have been through an ice age (Agassiz), the equally well-established fact that we are an intrinsic part of evolution (Darwin) and so on. These are some of the materialisations of the never-ending, conflict-ridden dynamics of knowledge advancement through the years. How this mainstreaming of new ideas comes about, we only have a limited understanding of as yet.

I know perfectly well that the so-called trickle-down perspective applied here has been heavily attacked by today's leading science historians, who, as a matter of fact, prefer to be called *knowledge historians*.[20] One after the other of these historians have questioned whether it is ever possible to identify the specific intellectual and geographical space where a certain piece of knowledge was produced by a certain ingenious individual at a certain time. They also state that it is impossible to distinguish the production of knowledge from its communication. Thus, they state, rather than being spread, knowledge circulates.[21]

As with many scientific 'turns', many important new findings have emanated from this 'circulation turn' in the history of science, in German described as a shift from *Wissenschaftsgeschichte* to *Wissensgeschichte*.[22] Perhaps, the most important ingredient of this shift is the identification of a lot of non-Western nodes of knowledge that have been advanced in interaction with the Western world wherein no node was subservient to the other.[23] It has also been observed many times that knowledge is not just received in a straightforward way but most often actively appropriated and, thus, adjusted to the needs and wants of the 'recipient'.[24]

Still, I think that the circulation perspective is overdone. For example, it is certainly true that cartographic knowledge in the seventeenth century was exchanged in a reasonably equal way between Chinese and French stakeholders. They were interdependent. However, it is as true that those who interacted in these matters were tiny elites, possessing and developing knowledge that only gradually spread outside the numerus clausus. In this case, knowledge historians have confused an outdated Eurocentric perspective on knowledge communication with a still adequate elite perspective.[25] This applies elsewhere too. For example, on the one hand it is now an established fact that Isaac Newton relied on a high number of rapporteurs worldwide for the development and establishment of his theory of gravitation.[26] On the other hand, it is as clear that it was Newton and not the rapporteurs who developed the theory in question, soon to spread all over the world. Thus, here it is possible to precisely identify a certain intellectual and geographical space where new knowledge was produced although that certain other intellectual processes had to take place before Newton's ideas were ripe. Such events are possible to identify in countless other cases too.[27]

In passing, it could be added that the history of science clearly shows that scientists of all ages have been aware of the need to carefully think through how to present their groundbreaking discoveries in order to make them attractive or at least acceptable to their contemporaries. Convinced as they were, they knew perfectly well that they had to be convincing too. Some applied a low-key approach, trying to tone down the novelty of their novelties in order to come across. Others applied the opposite strategy, declaring upfront to the world that their findings, pretended to be solid, really meant something radically new. Still others used a targeting approach, addressing only the most promising and open-minded people around them while saving intellectual energy by circumventing the stubborn conservatives of their time. Last, some were just nakedly honest, telling the truth about their achievements without concealing the probabilistic nature of their findings or any weakness that might be associated with them.

Copernicus and Galileo may be seen as examples of the low-key approach – Galileo, though, only at some stages in his troublesome career. Copernicus referred to 'Philolaus the Pythagorian (*c.* 470–385 BCE)' at least 'as an important precursor in proposing a moving earth [...] Even Galileo [...] repeatedly coupled Copernicus' name with that of Aristarchus of Samoa [...], to whom he (mistakenly) attributed the invention of heliocentrism'.[28]

Perhaps Joseph Lister could be seen as an exponent of a targeting approach. In opposition to most of his older fellow surgeons he introduced carbolic acid as prophylactic antiseptic treatment in surgery. My impression is that he directed his efforts mainly to younger surgeons who he considered relatively unprejudiced, finding it rather hopeless to change the minds among his older colleagues.[29] Finally, Charles Darwin may be regarded as an exponent of the honest approach, at least according to his own words in retrospect. In his autobiography he explains the almost immediate success of *The Origin of Species* by referring to his unusual habit to be as meticulous in considering counter-instances to his theory as observations which harmonised with it. Thanks to that, he anticipated many objections, which he responded to in advance, why he faced very few objections when his theory was made public.[30]

The state of the art and suggested steps forward

Over the last fifty years quite a few studies have been carried out on knowledge resistance, and some of them have also addressed ways to overcome it, not to mention a long tradition of intellectual history that maps minor and major steps forward in the scientific knowledge domain.[31] Thus, the research field is certainly no virgin land. In addition, in the over-crowded field of conformity and non-conformity research, there are also some relevant studies (see Chapter 3). In the 1970s psychologist Charles Lord and others demonstrated the presence of what they called 'confirmation bias', denoting people's aversion to change their early established beliefs, even when confronted with strong counter-instances. Today the concept is well established among social psychologists and other social scientists in the field.[32] Later on, a so-called disconfirmation bias was added. It signifies another impediment to knowledge acquisition, the empirically demonstrated tendency to put more effort into refuting other people's beliefs than one puts into scrutinizing one's own beliefs.[33] People seem to be as liberal towards themselves as they are tough towards others. Another discouraging but still provisional finding is the so-called backfire effect, whereby people tend to stick even more resolutely to their beliefs after having been informed that they are incorrect. Moreover, sometimes intellectually well-equipped individuals are better at sticking to false beliefs than their less well-equipped fellow travellers. However, recent studies seem to show that people who are intellectually well-trained have better chances, not only to distinguish false from true information, but also that they are more willing to accept what is true even if it runs counter to their ideological inclinations.[34] Obviously, more research is needed to settle the issue.

This is still not all. As has been shown quite recently, even when people are ready to adjust their perceptions when faced with correcting information (i.e. when the plan does not backfire in the narrow sense), it does not necessarily follow that they change their values and actions accordingly. Supporters of Marine Le Pen adhered to her even more strongly after having accepted that she was utterly wrong on a number of core issues. The same roughly applies to Americans intending to vote for Trump. Corrections succeeded, but the support for the Trump continued nevertheless.[35] Within this field, a number of experiments have also been carried out in order to explore ways to overcome knowledge resistance.[36] It should be added that not only emotional but also cognitive barriers to true knowledge have been addressed.[37]

Thus, over the years, a lot of scientific activities have taken place that aimed at gaining a deeper understanding of people's attitudes to knowledge, and a lot of insights have been gained too. Yet, there are still huge gaps to be filled in our knowledge on knowledge.

First, most studies on the matter are still rather small scale, and to date, they mostly concern conditions in the United States. In passing, over the last couple of years American researchers in the field have been almost obsessed with issues related to Donald Trump. It is however likely that they soon will recover from such a Trump bias – fully understandable as it is.

Second, almost no one applies a systematically comparative approach, which I find indispensable given that the mission is to reach a general understanding of the mechanisms at work that favour or disfavour the appropriation of true knowledge and not only a local understanding of these mechanisms.

Third, almost all researchers on knowledge resistance are basically oriented towards the present. If the past appears at all, it is the very recent past. So far, I have not come across any researcher applying a truly long-term perspective. That is why quite a few otherwise empirically serious scholars sometimes present-mindedly and carelessly state that our time is more knowledge resistant than other times. This is as poorly substantiated as the opposite view: that we live in a knowledge society. We neither know this nor possess the tools by which to get to know it. How are we to measure and compare levels of ignorance and enlightenment between the present and the past? Ideally an operational yardstick could be constructed so that we would be able to prospectively follow the destiny of knowledge, enabling future historians to study its trajectory in retrospect.

Altogether, social scientists and historians should make concerted efforts to move forward in our quest for a deeper understanding of the historically and culturally varying conditions by which knowledge is either appropriated or rejected and to move forward in our search for general patterns permeating variation. This means that historical investigations and observational studies should be combined with experimental studies of different sorts.

Among the profound questions to address are those mentioned in the introduction to this chapter. Are there certain types of knowledge that condition either its acceptance or its rejection due to either emotional or cognitive reasons? Emotional motivations could, of course, be of different kinds: ideological or personal and the like. The same applies to cognitive factors, ranging from the complexity of a certain bit of knowledge to the fatigue one may feel in face of information overload, irrespective of the complexity of each bit of knowledge. Many other circumstances may be relevant, such as the types of situations and cultural contexts wherein knowledge is to take root, the types of personalities addressed and so on. Last, the destiny of new knowledge may be closely related to the strategy applied by those individuals or minorities who represent it, depending on how counterintuitive and thought-provoking their message is, as well as on how much it challenges people's ways of living to a larger or lesser degree. A lot of other questions could of course be imagined, but these may suffice to give an idea of the nature of the contributions I would like to see.

Now, as I am writing these lines, the state of the art is about to improve considerably. Recently, a large-scale international research programme started researching knowledge resistance: its causes, consequences and possible cures.[38] However, this very promising and basically interdisciplinary programme, built around work packages on philosophy, psychology, political science and media studies, will not involve historical studies on the matter. I think that the inclusion of such studies would make the programme even better. Furthermore, since knowledge tend to progress through times, despite all kinds of resistance, the question should also be turned around: What explains knowledge breakthroughs among the citizens, and how do such transformations turn into knowledge-informed action? The investigation could be organised around an envisioned ideal trajectory, such as this:

- the knowledge gained
- knowledge resistance: cognitive resistance as well as resistance in practical life
- the cognitive breakthrough (i.e. when a significant proportion of the population accepts the new knowledge but still resists adjusting their actions accordingly)
- the practical breakthrough (i.e. when the internalised knowledge gains transform into real action on a large scale)
- a stage when these new action lines feed back as openness to another wave of knowledge gains.

As discussed earlier on in this chapter, such an approach could only be pursued by those who do not exclude the idea that at least some knowledge has an identifiable origin and that this knowledge, once produced, may spread to wider circles of people (although not unmodified).

Chapter 2

WHY DO SOME SOCIETAL PROCESSES AND PHENOMENA DEVELOP IN A CIRCULAR OR REPETITIVE WAY WHEREAS OTHER PROCESSES EVOLVE ALONG A CUMULATIVE TRAJECTORY?

Cultural evolution

Human interaction is uniquely dynamic in relation to what goes in other parts of the animal world, leading to incessant large- and small-scale transitions in our societies. This is due to an unrivalled ability among humans to intellectually handle what is not present, either because it is concealed or because it is absent from the actual setting. In turn, this ability depends on a similarly unique capacity for sequential thinking, not seen in other parts of the animal world.[1]

Through the ages, the magnitude of human interaction has grown immensely; today it is profoundly global in many respects. This also means that an ever-growing number of increasingly differentiated activities are increasingly coordinated, which is to say that human interaction is getting increasingly complex. This is so since, by my definition, the more coordination there is between the interactive parts in a system, the more complex it is – provided that other factors or circumstances remain the same. Or, differently put, given the level of coordination, the more differentiated it is, the more complex it is, as illustrated in Figure 1.[2]

It should be added that growing complexity often goes hand in hand with 'simplification'.[3]

These long-term transformations are culturally driven in the sense that they do not have any modifications of the human genome as preconditions, despite the fact that such modifications nevertheless do evolve, although at a slow pace in comparison to the pace of cultural change.[4] Of course, this does not mean that cultural change ever gets disconnected from or becomes

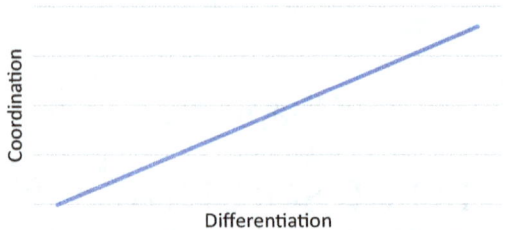

Figure 1. Complexity as a function of coordination and differentiation

independent of its peculiar genetic basis that we, as humans, embody. There would be no cultural evolution without the ever-ongoing 'investment' of our species-specific genome into the cultural system. Furthermore, it cannot be excluded that the gradual gene modifications that do take place may have an impact on the cultural fate of the human society, and probably have had such an impact. Still, cultural change would happen nevertheless, that is, with or without such an impact.

The dynamism of human society manifests itself in many different ways, partly as a process directed towards the differentiation of functions, habits and things, and partly as an oscillation between differentiation and homogenisation. When, in the distant past, early members of *Homo sapiens* took the first steps outside the Rift Valley in equatorial Africa to colonise most parts of the Eurasian continent, they had to adjust to hugely varying circumstances. And they did so, like almost no other species can do. Through this adaptive process, one and the same species unintentionally differentiated into many cultures. Thus, colonisation bred variation. However, as the process went on, the diverging cultures began to interact, which in the very long run made them less dissimilar than they had been earlier on when they only adjusted to varying environmental conditions.

This converging process was probably also largely unintentional, at least in its nascent stages. This, though, is not the end of the story. First, in a short- and even medium long-term perspective, the convergence of cultures might lead to increasing differentiation *within* them. Just think of all the recipes used for evening meals around the world. In a globalising world they tend to converge from one corner of the world to the other, while at the same time becoming richer and more differentiated in each place. A global reduction in the number of recipes from, let us say, 10,000,000 to 5,000,000 might be paralleled by a local rise from 500,000 to 1,000,000. If so, each culture would get not only a growing share of a declining repertoire, but also more recipes in absolute terms. From the bird's eye view, it looks like homogenisation, whereas from the frog's vantage point, it seems to be the opposite. Both are right.

Second, and more importantly, in a transitional process where cultural encounters gradually become more extensive, they tend to also temporarily become more conflictual. So, in the long run, what turns out to be the beginning of cultural convergence might, in the short run, appear as a series of irreconcilable cultural clashes, manifesting in widening (rather than narrowing) cultural friction. That such a rapprochement is silently advancing in the midst of great and sometimes violent turmoil is indicated by the fact that the clashes gradually tend to revolve around a common set of questions. It is thus likely that cultural convergence is initiated by a consensus – either implicit or explicit – on important matters of disagreement in order to eventually evolve into a reconciliation of specific attitudes.[5]

Describing the dynamism of human society in such general (and perhaps utopian) terms implies that history sometimes takes a cumulative and directed course and sometimes it seems to be moving in circles, although never really returning to its point of departure. It is apparently even the case that one-directional change (convergence) itself may be the driver of repetition (local redifferentiation). Above, these patterns are identified with a very high level of abstraction, except for the hypothetical example about eating habits. The question is whether they can still be visible if brought down to a slightly more concrete level. I think they can – both in relation to the directed changes and the recurring processes.

Cumulativeness

Humankind is an extraordinarily ingenious species whose history could be depicted as a multimillennial and countless chain of innovations. I think that by far most of those innovations are cumulative in nature. This means that their transmission, modification, renewal and occasional abandonment are the outcome of a multigenerational learning process. They are based on certain experiences which are deployed for further change, in the long run often, although not always, pushing history in a certain direction. This applies to the steadily widening networks of human interaction which today operate on a global scale, like the long-term establishing of sedentary life accompanied by a millennial growth in the number of humans inhabiting the world and the emergence of formally regulated human interaction. And so on.

These macro processes are based on innovations of many different kinds, and they are often also themselves intrinsic parts of such innovations, serving different types of needs. They concern all innovations, ranging from those primarily aimed at survival, through institutional constructs, as well as intellectual or existential innovations, to innovations that are made either just for joy or, on the contrary, for the destruction of joy through war technologies

32 BIG RESEARCH QUESTIONS ABOUT THE HUMAN CONDITION

and other devastating means.[6] Whether constructive or destructive, pleasant or unpleasant, beneficial or harmful, they are all manifestations of a uniquely human creativity and innovativeness.

It seems obvious that technological development evolves as a cumulative process, and insofar as this is the case, war technology is no exception to that. Apparently, the same goes for the advancement of intellectual innovations. The gradual development of writing systems, staff notation of musical compositions and mathematics are clear-cut cases in point, as illustrated in Picture 1.

It is perhaps less clear that the same applies to many institutional innovations, such as the development of bank systems, sports or the structuring principles of lawmaking. Almost all written law develops from initially being casuistically structured to gradually becoming systematised in an analytical and hierarchical way, where specific articles get subsumed under generally encompassing articles. Moreover, this happens independently in all kinds of cultural contexts. However, the evolution of the *content* of laws does not

Picture 1. Intellectual innovations

EVOLUTION OF SOCIETAL PROCESSES

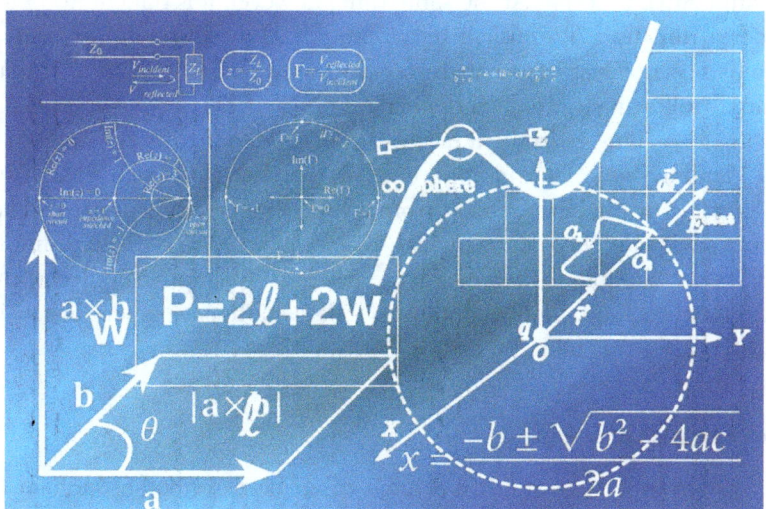

Picture 1. (*continued*)

necessarily follow a corresponding path, and a 'mature formal structure can sustain various and even contrasting moral attitudes, possibly implying that in the long run a mature form turns out to be more inert than the content of law'.[7]

What then about existential innovations aimed at coping with insight into our inexorable mortality, such as funeral rituals or other similarly sacred habits? Do they unfold in a cumulative way too? For instance, how should we designate the shift from pre-Christian cremation, through Christian earth burials to a secularised return to cremation? Were these changes based on intrinsic experiences of funerals where one initiative was the prerequisite for the next and so on? Or were they rather conditioned by ongoing ideological processes in a wider cultural context? This is difficult to determine. From a narrow perspective, these alterations look more repetitive than cumulative.

Anyway, as already said, directed cultural change seems to be cumulative in nature, which does not necessarily mean that cumulative change is always equal to directed change. For example, new musical trends do not emerge out of nothing; even highly experimental music appears to draw on less experimental ways of composing and performing music. Traces of romanticism and the Klezmer tradition could be found in the radically 'ironic' works of Curt Weill. Likewise, elements of classical and neoclassical structure can be discerned in the revolutionary compositions of Igor Stravinsky, such as the symphonic form. Doubtlessly, these creative manifestations are intrinsic parts of a cumulative process.[8] At the same time, however, it is difficult to pinpoint a general direction of musical development. How it should be portrayed I cannot say, not least since there are so many different musical traditions seemingly following mutually independent paths of development. This stands in contrast to some of the cultural processes mentioned above, such as the easily identified gradual and apparently irreversible shift from nomadism to sedentary life or the development from a low degree of functional specialisation to a steadily higher degree, which is closely linked to a gradual shift from a subsistence economy wherein a large majority were preoccupied with the procurement of food, to an economy of relative affluence where just as large a majority are doing all but that.[9]

Non-cumulativeness

Now, certain aspects of musical change could be described as circular or as a movement back and forth, although non-directionality is far from always equal to circularity. By momentarily bunching all kinds of music into one category, one gets the impression that musical compositions have been incessantly swinging from simplicity to complexity and back again. Of course, this is a somewhat simplistic view of musical development since such clustering does not do justice to a phenomenon that has followed so many different paths from different beginnings. Anyway, the possible recurrence of certain traits or aspects of music along its development, whether depicted as circularity or as a movement back and forth, applies to innumerable other phenomena along the trajectory of cultural change. Sometimes such movements are manifestly self-generating cyclical processes whereas genuine cyclicity is more often very difficult (if not impossible) to discern and demonstrate. Similar phenomena reappearing over and over are not necessarily the outcome of one and the same causal process; they may well result from different and even unrelated circumstances.

Whether intrinsic or not, such short-term fluctuations or long-term recurrences could be seen in most domains of human action as much in the development of ideas or ideology, as in social and economic development. One example of the former is intelligent design in the twenty-first century, very much resembling the so-called physico-theology of the eighteenth century.[10] Another example is xenophobia, which expands and shrinks, only to gather new momentum over and over again. There are obviously a lot of things that come and go. Perhaps the feverish fashion swings in most fields of consumption culture in modern societies are what first comes to one's mind. Another obvious example is the variation in economic activity, either conceptualised as short-term business cycles or as something that is claimed to evolve in long waves, each cycle spanning about fifty years. Other examples are the ups and downs in income distribution (see Chapter 5),[11] leniency and harshness replacing each other in legislation and courts of justice,[12] periods of political mobilisation and quiescence alternating with each other, war and peace,[13] and different aspects of societal stability and instability.

What explains the difference?

Despite the obvious difference between these two types of cultural change, both testify to their common independence of the slowly ongoing genetic modification of humankind. This is so since both circular and non-circular changes are carried out by generations of humans whose genetic differences are infinitesimal. But what explains the difference if it is not due to the genetic evolution of our species? Why are the cultural dynamics of human society enacted either as cumulative and often directed (and sometimes even one-linear) change or as recurrent and often non-cumulative change, whether described as ups and downs, movements back and forth or as circular?

A lot of research has been done on cyclical or circular and cumulative social processes. For example, theories and studies of the short- and long-term ups and downs in economic activity abound. Moreover, many of those theories (perhaps most of them) pretend to have identified a mechanism whereby those ups and downs are self-generating, which seems to imply that they are destined to repeat themselves infinitely. If this does not apply to John Maynard Keynes himself, it does indeed apply to Paul Samuelson's Keynesian theory of the business cycle.[14] The same applies to more loosely knit theories of long economic waves, once introduced by the Russian economist Nikolai Kondratieff and then reappearing in different guises ever since.[15] Even more loose are all the philosophies of eternal social repetition that have been presented over the past 2,000 years, which, by the late eighteenth century, were being replaced by ideas of the never-ending progress of human society.[16] Perhaps the Annales historian Fernand Braudel's once famous distinction between three different tempi of historical change could be seen as yet another example of this. He made a distinction between nearly immobile societal structures, depending on slowly changing environmental conditions, long-term 'conjunctures' and rapid political swings.[17]

However, against the backdrop of all the studies of either recurrences or non-recurrences, it is striking that so little has been done on what conditions some processes to evolve while others seem to revolve. As far as I am reasonably well-informed, today we do not have answers to these questions. But they are fruitful, inviting us to systematically inquire into the matter, but also enticing us to start formulating hypotheses.

The first hypothesis that comes to my mind is that innovations which turn out to be functional, or are perceived as functional, tend to propel cumulative change, that is, they foster a continuation of innovations along the chosen technical or institutional avenue. In contrast, cultural traits that are neither functional nor dysfunctional may come and go. Another possible explanation for the difference could be that cumulative novelties are a response to new either positive or negative experiences, whereas recurring phenomena revolve around what really are constant or recurring features and problems of human life and human needs. Or put slightly differently, because they roughly have an equal mix of merits and shortcomings, the same couple of political or institutional reforms tend to replace each other over and over again. Inherently emerging troubles with a certain promising package may trigger a return to a bygone but now equally promising package, which, once chosen, will lead to the same kind of troubles as last time [...] and so on. One example could be the vertical structure of certain organisations which, at some point, due to experiences of communicative deficit, get reorganised along horizontal lines, which in turn faces similar complaints of troublesome isolation between divisions or departments that lead to a return to the vertical structure and so on and so forth in a never-ending circle of organisational reforms.

Yet another idea would be that societies might oscillate between a backward- and a forward-looking cultural climate, in its turn conditioned by varying expectations of the future. In the former atmosphere, people may resort to established solutions or even bygone solutions, whereas in optimistic periods they tend to turn their back on the past in favour of everything that seems fresh and new. Societies can also oscillate between other positions, for instance, between collectivism and individualism.[18] Needless to say, all these forces and courses of action may be at work simultaneously, and there might of course be other forces at work as well.[19]

At first sight it seems obvious that technological change unfolds along a functional-cumulative-directional path, that is, in accordance with the first hypothesis suggested above. There are innumerable examples of this, such as telescopes and microscopes, which depend on the techniques developed earlier for blowing glass and so on.[20] One case in point could be the development of bicycle technology over the past 150 years or so. I imagine that upon the dawn of modern times, some people wanted to rationalise individual transportation that was still done physically, that is, without the aid of extra-somatic power (horse rides included but sailing excluded). The objective was to provide a device that was less energy consuming and that at the same time enabled a clearly better gear ration than pure walking or running. The two-wheeled vehicle was born. In its rudimentary stages it was just a velocipede – a so-called dandy-horse – without pedals, soon to be abandoned when the high-wheeler was introduced, a bi-pedalled cycle with a large front wheel and a small rear wheel, which was in its turn superseded by the safety bicycle, a chained vehicle with free-wheeling wheels of equal size and pneumatic tires (soon to be adopted by the car industry[21]) and eventually complemented by, although not yet replaced by, the modern multi-geared bike (see Picture 2).[22]

The example brings me to a couple of general reflections. First, this is a clear-cut case of cumulatively directed technical change wherein rudimentary solutions served as the basis for continuous refinement. Second, although bicycles have been exposed to almost incessant amendments and modifications, I know of no major technical change after the introduction of the multigeared bicycle technology, which happened many decades ago. This matter of fact invites the conclusion that basically bicycle technology is near perfection – the introduction of the electric bike can be considered a counter-instance. If so, it would explain the slowing down of technical change in the field of bipedal, physical transportation techniques.[23]

So, functionality breeds directed technical change, though this tends to slow down as the technology reaches a stage of (perceived) near perfection.

Now, as the repertoire of similar cases is almost endless, there are of course also a large number of far less clear cases. This is because technical change is far from always being purely technical and is often conditioned by a lot of non-technical needs, considerations and constraints, be they social, cultural, economic, political, cognitive and so on. Such constraints may contribute to

Picture 2. Bicycle technology from the eighteenth to the twenty-first century

Picture 2. (*continued*)

making technical development considerably less cumulatively directed than it otherwise would have been when not eradicating cumulativeness altogether, although this is far from necessarily the outcome of such ever-present 'impediments'.

In the book *Technological Choices*, published in the 1990s, a large number of such cases are presented. For instance, some seventy years ago chemical weed killers were introduced in some modernised, monocultural wine-growing districts of France, making the ploughing of vineyards pointless. The farmers continued with the ploughing nonetheless. Obviously, for some time they stuck to an obsolete technology for non-technical reasons, probably both for fear of becoming redundant and because of their ties to traditional ways of doing viticulture.[24] Another example concerns patterns of Neolithic technical choices and exchanges in the Jura Mountains 4,000–6,000 years ago. Whether more functional or not, culturally marked objects took far longer to be adopted than culturally unmarked ones. This was often even the case when the latter had to travel far longer distances than the former. For instance, around 2900 BC, people of the western Jura chose their axes from physically distant but culturally familiar communities instead of adopting them from nearby communities with which they were less familiar.[25] Furthermore, in many crafts a wide variety of techniques were applied, indicating a relative freedom of choice and a certain degree of independence from the physical or material conditions. Pottery is a good case in point, where one of the very few nearly ever-present moments is the rotation of the raw material, either through the movement of a turntable or by the hand of the craft worker.[26]

Thus, not only technological inertia but also technological change are often driven by non-technical considerations, possibly causing such change to be less than technically optimal.

However, none of this means that technology ever can be a completely free-wheeling activity, liberating its promoters from submitting to all natural constraints, such as the laws of physics or chemistry. This applies to pottery no less than to other crafts since the potters, applying very varied techniques, still have to adjust to a lot of physical circumstances, such as temperature and humidity. Furthermore, once established, a certain technical path dependency normally narrows down the options for further innovations, which is precisely a basic aspect of cumulative development. Moreover, the irrefutable fact that almost identical technologies often emerge in widely different cultural contexts clearly indicates that some technical steps are just technically determined and forceful enough to overrule whatever cultural roadblocks they meet on the way.[27]

It should be added that this apparent determinism is also visible in the field of institutional change, as it is in other non-technological domains as well.

Again, the development of the art of lawmaking can serve as a case in point since it has developed from being casuistic to being analytically hierarchical and thus increasingly encompassing, regardless of the peculiar societal setting wherein it evolved.[28] Indeed, legislation could also be considered a technique. However, in my discussion this does not make any difference since the point to be made is not primarily about technology but about cumulativeness.

So, here I stand with a grand, vital and largely unanswered question about the forces explaining why some societal processes evolve in a cumulative way whereas others do not. But how can it be properly addressed in accordance with the overall criteria presented in Part 1? Right now, I have no clear answer to this question, but I would be very interested to know the reader's ideas on the matter.

Chapter 3

WHY DO SOCIAL NORMS CHANGE, DESPITE THE FACT THAT THEIR MISSION IS TO BE SUSTAINED? WHAT ROLE DO NON-CONFORMIST INDIVIDUALS AND MINORITY GROUPS PLAY IN CULTURAL, COGNITIVE AND NORMATIVE CHANGE?

Above, the reader finds a heading consisting of two questions. This is to stress that they are intimately related and deserve to be approached together. At the same time, however, they are separable, both as such and because the second question has developed into a large research domain in its own right over the years. So, the two questions will be treated in juxtaposition, one after the other. Moreover, the second question will be discussed in some detail whereas the first one will be only briefly touched upon, simply due to the fact that I have familiarised myself more with the second question.

A: Why do social norms change, despite the fact that their mission is to be sustained?

Generally, human interaction is regulated – if not from the very outset in the development of all societies, at some early stage. Such regulation is enacted both by means of physical power and with the aid of norms. But to a substantial degree, physical power also relies on something other than itself since it would not function without loyalty from strategic groups and ultimately from the subjects or citizens in general. Norms serve to stabilise social interaction, be it peaceful or characterised by more or less violent conflicts. If such stabilising measures are successful, they will strengthen the norms through a series of feedback loops. Norms are introduced in order to be preserved through vertical and horizontal transmission, and thus, they promote human action in accordance with the norms. But they change, nevertheless, some more than others and more so in certain situations and under certain circumstances – even those that appear especially stable, sometimes even timeless. Some become stricter vis-à-vis human needs and desires whereas others get more relaxed. According to a new study of meta norms in 57 countries, in modern, basically liberal countries, what is called *spreading of information* is the most typical informal sanction applied, whereas verbal confrontation and avoidance of norm breakers are more frequent in more hierarchical and less liberal countries.[1] Since modernisation is a historical process, this clearly indicates that norms do change.

On the micro level, change of attitudes towards lethal violence as well as non-lethal violence are paradigmatic examples of such long-term normative shifts. It goes without saying that both animals and humans largely try to avoid getting exposed to lethal violence, which nonetheless endlessly continues to strike unlucky members of all species. But whereas violence among animals shows no tendency to either rise or fall, homicide has gone through an undeniable long-term decline over several centuries, despite certain swings up and down.[2] It is beyond reasonable doubt that this decline applies both to interpersonal violence and to violence in social conflicts (although right now we seem to witness a growth of violence in such conflicts), while it is so far less certain that inter-state violence has experienced the same downward trend.[3]

Norm change is also linked to the decline of non-lethal violence. For example, in the very long run it has become less and less acceptable to expose children to violence from adults. This has also meant that parents and other adults must negotiate with children early in the children's lives, often by resorting to rational arguments. This may have contributed to the generally improved sequential thinking among children and young people, indicated by, among other things, the so-called Flynn effect.

Moreover, it also appears obvious that such trends display important norm changes. It is, however, for that reason not clear whether it is a matter of change towards increased constraints on human needs and desires or change towards reduced constraints. It is rather both, as seen from the perspective of the increasingly curtailed perpetrator and, at the same time, from the perspective of the potential victim, gradually being released from fear.

This double-edged norm change is also visible on the macro level of societal change. Here one could mention the fact that in all countries of the contemporary world, one finds an extensive central administration, a tax system and encompassing welfare institutions, as well as a judicial system granting the citizens at least some inalienable human rights. This seems to be the outcome of a secular process leading both to centralisation and to decentralisation in many countries, for example, in the form of individual human rights. Through taxes, and by other means, the citizens contribute to a set of collective solutions to a high number of social and economic problems that were met with other means (if at all) earlier in history. Through certain laws, citizens are granted individual autonomy in areas of human interaction where in the past they were exposed to disciplining control by the authorities. This is how the trajectory of societal change appears in many societies. Looked upon in this way, societal change seems to be a continuous balancing act between collective and individual solutions to problems in human interaction. Perhaps, it is even the case that individualism and collectivism can only evolve in concert. Or are they rather in constant opposition to each other? This needs to be systematically addressed, as do the different outcomes of various ways of reconciling individualism and collectivism. But whatever would be the result of such inquiries, changes such as these are intrinsically normative.

How are the dynamics of such normative change to be explained? And why are some norms more stable than others? These are the questions to be asked. One might be inclined to suggest that normative change is due to three general experiences: reliable threats against the sustainability of life in accordance with present norms; reliable expectations of substantial improvements to life conditions thanks to substantial norm change; and unnoticed norm change due to a gradual change of ways of living. Being intuitively true, these notions are still far too unspecific and, so far, they are also just a set of hypotheses still to be operationalised and tested.[4] However, both the ongoing climate crisis and the corona pandemic point in this direction.

Many suggestions as to how to explain these processes have been offered, but so far there have been none that I find satisfying. According to one idea, norm change is driven by 'my way' personalities, that is, by people who get more disturbed by certain unwanted behaviour than others do and intervene through punishing reactions, or next-to-punishing reactions, towards people who indulge in such behaviour. This will start an evolutionary process whereby, step by step, more strict norms get established. Long-term declining interpersonal violence and people's gradually inhibited spitting habits have been put forward as manifestations of such long-term cultural processes.

Whereas this theory may reasonably well explain growing strictness, it has little, if anything, to say about cultural processes leading towards increased tolerance of certain human needs or desires. It might explain the declining tolerance of violence, but not the rising tolerance of people's sexual preferences. A better approximation would be to identify what might be common behind these seemingly contrasting changes, namely that both stem from a growing insight into other people's needs.[5] But, such an explanation remains to be demonstrated too. In order to get closer to convincing and reliable explanations, a lot of comparative historical research is needed. In a recent article, mathematician and social scientist Pontus Strimling and others are presenting and testing a theory, according to which change towards increased tolerance depends on values, such as fairness and avoidance of harm, that most people share, be they so-called liberals or conservatives.[6]

One could approach the question from another angle by asking what makes a society hold together and why it does not fall apart due to all the conflicting interests haunting it. Or is it rather the case that conflicts between social classes and other groups promote the development and unity of society, and norm change? These are questions that have been asked many times in different ways. However, they are not often addressed together. How are they linked? Perhaps it is precisely through the struggle between people with vested interests that societal coherence evolves and is sustained. Provided that this is the case, it may be linked to growing societal complexity (i.e. coordination of a gradually growing differentiation of social functions) and probably also to the long-term decline in physical violence. These questions are of paramount importance for our understanding of the structure of human society.

B: What role do non-conformist individuals and minority groups play for cultural, cognitive and normative change?

The question – an introduction

Being an extension of question 3A above, the following discussion could have been integrated there as one aspect of the overall question. But since I, in relation to this specific question, have come across a big and extensively researched and rather independent scientific domain, I find it more convenient to treat the question in a separate section.

In all times and all societies there exist certain minorities as well as autonomous individuals who, through their new and original ideas, dissociate themselves from the dominating beliefs and thoughts of their time, as these are represented either by the majority surrounding them or by power-exercising minorities. It is natural that initially such ideas are met with discomfort or even outright hostility, more so the more counter-intuitive they are and the more they challenge people's way of living. At the same time, it is quite common that majorities of any time praise those minorities of the past that eventually have come to represent their own currently dominating views. This may indicate a wish among the conformist majorities to be more non-conformist than they are or have the courage to be but does also indicate that those minorities probably play a pivotal role for the unique cultural dynamics of the human society. Many circumstances suggest that minority groups that are highly dedicated about their cause have a larger impact on the development than their insignificant number might indicate.[7] But the cognitive and normative significance of such groups for cultural change is still unsettled. Surely, there is a long-standing tradition of research on conformity and non-conformity, but there is a lack of knowledge about this particular question. Likewise, to my knowledge, very little research has been done about the risks some people have been willing to take to bring about certain societal changes that otherwise would not have occurred and for which they, given that they eventually do materialize, will get praised only in retrospect.

In the following I will go into detail about the state of art in research fields that are relevant for these questions and present an outline of a research design as how they could be addressed.

State of art

One essential part of such a design would be the relationship between the pressure to conform to local or "global" majorities and the counter-pressure they are sometimes confronted with from dedicated minorities. This includes, first, the risks some people have been willing to take, the discomforts they have endured for the sake of their desire or inner compulsion to hold fast to their convictions and their chosen course of action. Second, the strength of the pressure not to rebel that is exerted by dominant groups should be studied. Finally, each of these two forces should be studied in relation to the other so that pressure and counter-pressure could be evaluated.[8] The nature of that relationship, now and in the past, is not self-evident and calls out for research.

There is a long tradition of research on conformity within psychology and social psychology. Certainly, the focus from the beginning was mainly on why people do *not* resist group pressure, with the much-discussed studies of obedience and authority by Solomon Asch, Stanley Milgram and Philip Zimbardo as perhaps the most well-known today.[9] In particular, the studies performed by Asch in the 1950s constitute a strong foundation for research on conformity and its opposite.[10] In turn, his research was an offshoot of Muzafer Sherif's research in the 1930s on conformity with social norms – conformity that appeared to be depressingly widespread.[11] This was partly what inspired Asch to carry out new studies, under different conditions and in such a way that subjects were asked to take a position on clearly right and wrong answers. It was thought that this would make it easier for subjects to take independent positions,[12] but it did not help: conformity was still pervasive, as was the researcher's disappointment at the state of affairs. Or as Asch expressed his distress: 'That we have found the tendency to conformity in our society so strong that reasonably intelligent and well-meaning young people are willing to call White Black is a matter of concern.'[13]

Early on, a number of subtle distinctions were sown in this field of research. Scholars began to differentiate between informational conformity and normative conformity (1955)[14] and between internalization and mere compliance (1953).[15] This conceptual development, arising partly from empirical evidence, may have contributed to the increasing focus of research on why some people, despite everything, resist the pressure to conform.[16] There are now numerous studies on both the dominant inclination to go along with the majority and the less common but always present inclination not to do so – as well as a sometimes heated, even rancorous, discussion among proponents of differing opinions on the matter. There are thus many results to take a position on here.[17]

These include psychological research that fancies it sees a change towards diminishing human opportunism, but also studies that seem to show that no such evolution has occurred.[18] Such studies, however, reach back only to the 1950s, while studies that relativise the present day from a long temporal perspective leave much to be desired.[19] The standard experiments have been carried out on male American students,[20] often students of psychology at the time and at the university where the researcher worked.[21] Even though generalisations from the individual experiment to the universally human are bound to stumble as long as historical and cultural variance are ignored, such generalisations are nevertheless fairly common, although there are also several studies that attempt to trace the significance of cultural variance as an explanation of differences in conformity with majorities.[22] An oft-repeated conclusion is that people are more conformist in collectivist cultures than in individualist cultures[23] Another observation is that people in small-scale communities are more pressured to yield to the will of the local majority than are people in large-scale communities.[24]

One possible problem with most of these studies is that they are based on an overinterpretation of the findings of Asch and other researchers of conformity. This is at any rate what Bert Hodges and Anne Geyer assert in an intriguing article, where they challenge both the interpretations and the disappointed evaluations of the outcomes.[25] On the one hand, they argue that the conformity among the subjects of Asch's studies was not as pronounced as most interpreters of his experiments have claimed. For example, as many as 75 per cent of subjects agreed only three times out of twelve with the intentionally wrong answers of confederates in a line judgement task (a type of study that has been repeated ad infinitum).[26] On the other hand, Hodges and Geyer argue that people may have good reason to sometimes submit to the opinion of a local majority even though they realise it is objectively wrong. In so doing, they argue, people are often pragmatic rather than conformist, oriented towards finding solutions rather than protecting their image, sensitive rather than cowardly.

They base this on the notion that people have a 'heterarchy' of values; that is, they agree with different and sometimes competing norms or values that are not in any fixed, hierarchical order. Only one of these norms is based on maintaining one's position, while a second has to do with keeping the group together, and a third with not embarrassing other people, and so on. Those who are seeking to gain majority support for important values cannot do so willy-nilly. 'We pick our battles,' as they say. But in spite of this, a larger percentage of subjects have exhibited non-conformist behaviour in such experiments than the conventional psychological wisdom claims.

The perspective that Hodges and Geyer adopt aligns rather neatly with the "interactionism" suggested by Stephen Reicher and Alexander Haslam, based on rejecting one of the conclusions of the conformity studies of the 1950s, that the group threatens the individual's autonomy.[27]

Judging by the highly reductionist overview above, I can conclude that psychological and social psychological research on conformity and non-conformity is incessantly in flux. This is surely to be expected, for what in the world is not constantly changing? Not least importantly, there is now a plethora of studies of cultural differences in conformity – and yet, psychological research is, understandably, relatively void of history. At the same time, there is a paucity of historically oriented research devoted to this question, even though it has always been and will always be of profound importance to human social development.

Naturally, historical presentations are replete with heroes and anti-heroes who have battled against the odds – and countless studies of the odds themselves, the attempts of the more or less authoritarian governments to quash or neutralise rebellion. Case studies of particular events when people yielded to destructive group pressure also exist, of course, some with references specifically to the psychological research on conformity. One example is historian Christopher Browning's work on German reserve police ordered to help exterminate Polish Jews in the summer of 1942. An overwhelming majority of the police men chose to participate in the massacre, even though the commanders offered them the choice to opt out of the duty with no repercussions. Browning looks for the explanation in the work of experimental psychologists like Stanley Milgram and Philip Zimbardo,[28] that is, in the idea of 'evil' as banal and situational, rather than conditioned by particular traits in certain individuals.[29] This understanding seeks its origins in Hanna Arendt's conclusions about the trial of Adolf Eichmann, an understanding she summarised thusly in her book about the trial:[30] 'The trouble with Eichmann was precisely that so many were like him, and that the many were neither perverted nor sadistic, that they were, and still are, terribly and terrifyingly normal.'[31]

There is much value to Browning's study, even though researchers of conformity have recently challenged its theoretical basis, both Arendt's trivialisation of Eichmann's motives and the "depersonification" of "evil" that guided the work of Zimbardo, Milgram and others.[32]

Anthropological and evolutionary research also comes into play, among else in studies of differences in conformity across ethnic groups.[33] But as yet there are no studies that attempt, based on comparative and empirical findings on long-term historical variation, to seek out possible generalisations concerning the tendency to go along or go it alone when it is impossible to do both. And thus far I have not found any experimental or empirical-historical studies of the relationship between pressure and counter-pressure. For sure, there are small, isolated studies with some kinship to my questions, such as the reflection that the unexpectedly strong conformity found in Asch's 1950s study can be partially explained by the authoritarian pressure of McCarthyism,[34] or a study of the few who took extreme risks when they saved Jews from the Nazis.[35] But, as said, these studies are small and not systematically performed.

One type of study that I envision concerns the thinner slice of the question: that which unites or does not unite the minority who, despite the risk of having to endure severe hardships for maintaining their positions in word and deed, have nevertheless done so.[36] This encompasses studying the risk as such; that is, how the strength and nature of the pressure to conform have varied across time and across cultures, which includes studying a number of macro conditions as well as a large number of individuals who represent the long historical sequence of resistance by the very few. I will explore several aspects of this below.

How do people manage the magic trick of simultaneously adhering to their convictions and being liked – as long as they do not spontaneously agree with everybody else, or belong to the minority who can manage it with a kind of artfully artless ease? To put an even sharper point on it: how do they find the courage when the risk is much greater than simply being disliked by others? The question can be scaled up to two attitudes towards laws, norms and violation of norms: obeying the law above all or obeying one's convictions above all.[37]

The first attitude is that we should obey the law, whether or not we live in a democratic society under the rule of law. According to some research, the primacy of obedience is most pervasive in collectivist cultures – democratic or not (apart from regimes of terror where there is no choice at all).[38] The minority (everyone) has a duty to submit to that which the majority (or the government) has decided. This applies first and foremost to acts, but for some also to attitudes most pervasive in collectivist cultures – for many extended beyond the written laws to the unwritten. This could be called conformity to group pressure on the macro level, or conformity to the 'will of the public'.[39] And it also seems as if most people obey most laws most of the time. This makes sense – both that people obey the law most of the time and yet do not obey all of the time. If no one or almost no one obeyed the law, the law would be a dead letter, and if everyone did so, the law would be virtually superfluous. The existence of the law (the courts and the penal system) implies anticipation that laws will be broken.[40]

A second attitude is based on the idea that we should, above all, obey our convictions – in compliance or defiance of the law, with or against the majority. Or as said about the Swiss border police commander Grueninger who defied the law to help Jewish refugees enter Switzerland after the German *Anschluss* of Austria in 1938: 'People came first and the law came second.'[41]

To a certain extent, one finds both attitudes held by the same people, and some who mainly live according to the first perhaps wish that they lived more according to the second. One expression of this is all the posthumous homage to heroes who resisted dominant norms and 'accepted truths' in their own time and were later vindicated according to contemporary mores. But contemporary mores are something other than the judgements of the many or the majority. The many who usually pay homage to those who resisted group pressure in the past and were vindicated with time may also be the same people who refer to contemporary majorities in support of their positions on current issues. To put it another way: autonomous individuals are admired for their resistance to past majorities, but only once the former have gained the support of the contemporary majority. And they are often admired despite – or perhaps because of – the fact that in their stubbornness they sometimes felt compelled to break the law. This can be regarded as a sadly anxious *ersatz* autonomy, opportunism with anti-opportunistic attributes. But in this admiration one can also see a genuine but abortive yearning for the courage of conviction, if not occasionally more of an inability rather than unwillingness to see beyond one's own culture. In any case, this sheds light on the changes in group pressure itself, both in its strength and with regard to its object and content.

Generally speaking, some of the norm-breakers of the past are revered, those whose convictions and conduct align with contemporary norms. And this occurs timelessly in every present – our own and that of others. This seems fairly safe for the choir singing songs of praise, whose tones rise and fall, keeping time with the spirit of the times. If the heroes of the past were not really as pure as contemporary observers would have them, it is not unusual for the anomalies to be erased or modified.

One might say that every generation produces a minority of independent people whom a later generation of people ennobles – ennobling themselves in the process. And thus, they make the heroes of the past more contemporaneous with themselves by touching up the picture, erasing the flaws that would put them in the minority of the new generation – but perhaps not of the generation to which they belonged. Mythologisation, as such is called.

In terms of social development, however, the most interesting aspect here is that such minorities are an utterly natural part of the never-ending cultural dynamic that characterises the human society – even if these people far from always had any intention to generate social change. It may be that many were simply incapable of walking away from their conviction or their obsession, regardless or in spite of the consequences. However that may be, they personify the eternal tension between preservation and rejuvenation, the hallmark of cultural evolution.

The significance of outsiders for cultural evolution

Cultural evolution is impossible without transmission across and within generations. But it also requires rejuvenation in the form of innovations of many kinds: moral, institutional and technical rejuvenation which is, in turn, transmitted. What significance do the non-conformists have as the drivers who ensure that cumulative cultural developments carry the unique dynamic that characterises human society? A minority of oppositional figures is underdogs of a sort, but by 'standing up for their beliefs' these people can often have an influence that is proportionately greater than their percentage of the group or the population.[42] They are akin to those whom Arnold Toynbee in his time called the "creative minority", who must always remain a minority even though every individual's life contains elements of creative problem-solving.[43]

Things are rarely quite this simple, and when it happens, it rarely happens without resistance and pain. Some human biologists also believe that conformity is adaptive, that is to say both biologically and culturally beneficial, in the sense that the dominant values of a group seem to be those that keep the group alive. It is therefore wise to submit to the group. They argue that conformity is rewarded, socially and biologically, because "humans are 'programmed' to conform".[44] They do not however deny that such conformity may be distorted and maladaptive and would probably not either deny that evolution could be 'straightened out' by the small group that sees through what is called "frequency dependent bias" in the large group.[45]

A possible design

This in turn leads back to the questions I think could guide a study of resistance to group pressure.

To identify resistance to group pressure, we must also identify the pressure. And such has been done in all research on conformity and its opposite. Yet, there are no studies of the dynamic interplay or dynamic tension between the two, and particularly not over the long term. For example, I have found no studies specifically preoccupied with the risks people are prepared to take in order to maintain their convictions or their chosen course of action, and it is precisely such a study that would be able to shed light on this tension.

With this in mind, a possible study should encompass the question of the universal and the question of the historical and cultural variation in pressure, counterpressure and the relationship between the two. To answer such research questions, all of this must be studied historically and cross-culturally.

Three questions about the universal or the timeless come to mind. The first is whether there is anything that socially and psychologically unites the minority of people who in the most variegated situations resist the pressure of a majority, either a local majority – those with whom the resisters interact most closely – or a 'global' majority, such as the majority in their country of residence. As said, this does not mean that I want to identify the heroes of history and society. Those who resist the pressure might be anyone at all: people of strong moral convictions, rigid people incapable of change, authoritarian people who cannot abide other authorities, people with strong faith in their own knowledge and so on. The second question has to do with what might be common to the pressure to which people have been subjected. The third question involves finding out whether there is any general pattern for what kinds of risks people have been willing to take in the force field between external and internal (counter) pressure.

The same questions may be asked about the time-bound and the culture-bound in this push-and-pull relationship. In turn, they can be broken down into additional question complexes.

Based on the foregoing, I would suggest a future study focused on the relationship between the social imperative to conform to the norms of the majority or the government and the resistance that has been proffered these norms. I am especially intrigued by the risks people are and have been willing to take, given that they have already identified with a minority position, and I foresee that in a study of a cultural process like this a long temporal perspective will be applied.

So such study should encompass:

- Macro historical analyses of pressure and counter-pressure
- A series of analyses of individuals across a long span of time
- Contemporary experiments in which scenarios of risk, pressure and counter-pressure are simulated

By this, the overarching question of the evolutionary significance of dedicated minorities may also be analysed and specified.

Chapter 4

DOES A GRADUAL EXTENSION OF OUR LIFESPAN (AND THE RISE OF WELFARE) IMPLY A GROWING OR DECLINING ABILITY TO POSTPONE THE SATISFACTION OF OUR NEEDS AND DESIRES?

The relevance and importance of the question

With this chapter I would like to encourage humanities scholars to step up their intellectual efforts to understand the societal significance of people's ability to postpone the fulfilment of their needs and desires – except the sexual ones (see appendix B). Optimally, such a task should be a concerted undertaking by historians, psychologists, sociologists and others to advance our knowledge of both the long-term destiny of this faculty and the mechanisms behind its varying presence among different peoples and cultures.

The ability to postpone rewards or the fulfilment of needs and desires is essential for all human development – both at the individual and societal level. Otherwise, we would not survive, and no society would cohere without some balance between the fulfilment and postponement or suspension of some of our cravings or even definite abstention from some of our cravings. As stated by Roger Scruton in a little book about beauty, 'desire […] is inherently bound by prohibitions'.[1]

For the sake of our personal interests, we already have good reasons to resist and postpone some of our needs, either to forget them once and for all or to fulfil them at a point in the future when the opportunity cost of doing so is as low as possible. We only live once; true, but all our days, from the first to the last, are parts of that life – why each one of them deserve to be equally valued and enjoyed. Let alone the fact that such an even distribution of pleasure will never come true. In passing, the temptation to procrastinate (i.e. to delay what should be done at once) is just another aspect of the issue at stake in this chapter.

Furthermore, since most resources are scarce in relation to their overall attractiveness, all of us also have good reason to step back sometimes for the sake of other people's desires. Clearly, such deliberate willingness to suppress one's immediate impulses for the benefit of other people's access to much-wanted material or immaterial resources implies the presence of some sort of empathy among humans. Anyway, whatever the neurological mechanism behind this empathic ability, it is obviously in the interest of each of us to apply it or, as expressed over and over again from the most ancient times until today, to do to others what we expect them to do to us (see appendix B). We have a mutual interest in empathy. The Golden Rule is one of its manifestations.

Now, since society can be conceived of as the sustained regulated interaction between human individuals, what applies on the individual or micro level also generally applies on the societal or macro level. Yet not without certain Keynesian modifications, meaning that under specific economic circumstances it could be as rational for an individual household to save as for a government to spend. A society is also the machinery for the regulated production, preservation and consumption of all kinds of resources that are allocated between both its current members and those yet to be born. Such societal machinery would not be a society if its agents were generally unwilling to trade instant rewards for future rewards, nor if current needs were completely discounted. A society whose members were incapable of thinking and planning beyond the present would hardly be a society at all. It would have no continuation – it would be just a theoretical construct. However, whereas history abounds in societies, perhaps predominantly empires, that have collapsed under the weight of their short-sightedly over-consuming elites, I do not know of societal breakdowns due to excessive hoarding. However, collapse due to hoarding among the privileged at the expense of their oppressed subjects in the present is commonplace.[2]

Today, almost every citizen is far better off than even most of the privileged citizen of the past. Robert Fogel also claims that between the beginning of the eighteenth century and late-twentieth century, the peoples of Europe seem to have experienced a clear reduction in the inequality of income distribution.[3] This, though, is disputed, which I will discuss in more detail in Chapter 5. Mass consumption has not closed income gaps, just moved them to a higher general standard of living.[4] Anyway, people live longer and have more healthy lives, and in our times of mass consumption, they have access to a plethora of things that only the richest of the rich could afford in the pre-welfare era.[5] Or as expressed by American economic historian Robert Fogel: 'Today, ordinary people wish to use their liberated time to buy those amenities of life that only the rich could afford in abundance a century ago.'[6] In addition, people of today are exposed to innumerable enticing joys and to things not yet possessed that were probably far beyond the imagination of past generations. This is the case despite the fact that the economic gaps over the last couple of decades have widened considerably in most modern societies.[7]

Two coupled historical processes lie behind this development. The first process is the industrial revolution, enabling an escape from the Malthusian checks which, according to Robert Malthus, had haunted humankind since times immemorial and would have continued to do so until the end of days. One important factor behind the breakout from this vicious circle of repeated demographic collapses was the introduction of fossil fuels. It included a transition from a somatic energy regime to an exosomatic energy regime. Over the long run it has not only led to an era of sustained material economic growth but also led to a release from physical suffering among millions of industrial workers (in the West, nota bene!). However, in its nascent stages, industrial capitalism made living conditions even worse in its core areas, in particular for the workers in unhealthy industries but also for all urban citizens exposed to heavy pollution.[8]

The second process was the gradual emergence of a welfare society. In many countries it was accompanied by the establishment of a social welfare system, encouraging people to fulfil their individual life plans more than ever before in the history of humankind. Thus, collectivism and individualism went hand in hand – paradoxical as it may seem.

Now, the release from the gloomy predictions of a poor static society did not last forever, though many enlightened historians believed it would when they looked back from a vantage point of the 1960s or early 1970s.[9] It is precisely those fossil fuels which, towards the end of days, are responsible for the re-emergence of the Malthusian checks that many believed humankind had emancipated itself from once and for all. We are about to be entrapped by the same process that set us free in the past.

Furthermore, the combined outcome of economic growth and social welfare systems have doubtlessly made life a lot easier for millions of people – life has become freer and at the same time safer. On the other hand, the consumption culture that ensued may also have disinclined people from feeling responsible for the long-term preservation of the material basis of the human civilisation. Without any checks on and balance of energy-demanding private consumption, such behavioural patterns may amplify the impending crisis of our civilisation.

So, what once worked as liberating forces for humanity may in the long run transform into fatal threats to humanity.

Against this background it is urgent to ask whether the people of today are less willing to delay gratification than any previous generation. Are contemporary citizens more impatient, more self-occupied or 'narcissistic' than people have ever been, as has been put more than once since Christopher Lasch published his influential work on the culture of narcissism in our time in the late 1970s?[10] Are affluent people more short-sighted than less affluent people? If so, does this apply on the macro level too? Human geographer Daniel Dorling's book *Injustice* is loaded with examples pointing in this direction, such as new cars, clothes, ever bigger homes and 'the purchasing of new looks for ourselves'.[11] Dorling points out that 'times of great inequality are typified by great excesses'.[12] He is not the first to do so.[13] Also, political denial of the human contribution to climate change or the inability to make political decisions in accordance with the lip service paid to such matters seem to indicate short-sighted over-consumption, together with a lot of other day-to-day decisions typical of contemporary world politics. The extremely hot summer of 2018 is a case in point. It certainly evoked a widespread fear of an impending human-made catastrophe for life on earth, but soon waned as the weather seemed to normalise again. If the disastrous fires in Australia 2019–20 will have any lasting impact on people's attitudes, remains to be seen.

The fact that many things which required decades or even centuries for their completion in the past can now be executed in a few years may point in the same direction. For example, it took about a hundred years to build St Peter's Basilica in Rome whereas similarly complex buildings of today may be erected ten times as fast, or even faster.[14] Thus, engineers engaged at the beginning of such huge building projects of the Renaissance could not even expect delayed rewards for their initial intellectual investment, not to speak of the workers sustaining all kinds of pain without getting very much in return. Today, most of those involved in such projects, from decision makers to construction workers, have a good chance to cash in on the rewards rather quickly one way or another. So, in a time when life was short, in many cases people had to wait way longer to get rewarded for their efforts than the long-living people of the contemporary world. Such a material need for the sustained sacrifice of fulfilment might have had as positive effect on the ability to sustain it as the relaxing of this need among the citizens of today might have a negative effect.

But this remains to be substantiated. Perhaps, it is rather the case that the short-sighted discounting of future rewards is our common, timeless fate as humans. This we do not yet know, but it is of paramount importance to gather knowledge about the changing conditions regarding people's willingness and ability to strike a reasonable balance between immediate and future gratification.

Now, has this question not already been on the research agenda for many, many years? What remains to be done? Have all such questions not already been asked over and over, and all the possible answers been given as many times?

The state of the art

Certainly, the disciplining of manners is a far from new endeavour among researchers. From different angles and disciplinary points of departure, various aspects of the matter have been addressed many times before by social scientists and by humanities scholars. Among the latter it is perhaps mainly historians who have approached the question. Below I give some glimpses of what has been accomplished so far. It has to be seriously incomplete, partly because of the limited ambitions of this book, but also partly because the question relates to so many fields of research that a complete overview would be difficult to provide. I hope that others with specific expertise will step in and fill out the picture.

Discounting between consumption today and consumption tomorrow, also called *intertemporal choice*, has been at the core of economics for a very long time, empirically no less than theoretically. Since such discounting is precisely about postponement, it is relevant for the question at stake in this chapter. According to conventional economic theory, people's preferences are rather time consistent, so the same should apply to their discount rates. In later years this assumption has been frequently questioned, bringing economists to recognise what to others appears rather intuitive, namely that 'people sometimes behave as if there is no tomorrow and at other times seem obsessively focused on the future'.[15]

One strong observation in line with this is that people normally have declining discount rates, showing that they are not fully aware of their future degree of self-control. For example, being naively overoptimistic about one's ability not to procrastinate in the future makes procrastination in each 'now' more likely.[16] A related observation is that entrepreneurs tend to overvalue their long-term resilience. On average, they think that 60 per cent of small-scale firms will survive more than five years from their establishment, whereas in fact no more about 35 per cent do so.[17] This certainly does not directly reflect people's illusions about their self-constraint but it exposes a naïve conception of the future.

There are a lot of similar observations which I cannot go into detail about here.[18] Let me just point to the overall conclusion drawn from such observations: it may be more fruitful to attempt a better understanding of the psychological processes behind unstable preference scales than to develop better discount functions.[19]

Like economists, psychologists have always been interested in the tension and balance between short-sighted pleasures and long-term fulfilment. But distinct from economists, psychologists neither limit themselves to issues related to the consumption of commodities and services in the conventional sense, nor do they often conceptualise and calculate the trade-off in terms of discount rates. Instead, they talk about self-restraint or self-control, and not only in the economic spheres of life, narrowly defined, but in all sorts of human affairs – disregarding whether it is possible to subject 'non-economic' affairs to such calculations. Moreover, whereas economists tend to apply a theoretical top-down approach, psychologists normally begin with the phenomenon per se and start theorising in the light of experimental findings.[20] However, despite the very different ways of phrasing the problem, it is roughly the same, and it is one of the defining themes in psychology and economics.

Self-restraint has been addressed over and over, as much by cognitively oriented psychologists as by psycho-dynamically oriented psychologists. For example, a central part of Jean Piaget's developmental psychology regards the child's mental maturation from egocentrism towards de-centrism, that is, from partly self-blind self-occupation towards having an ability to also see things from the other's point of view.[21] In psychodynamic theory, as developed by Sigmund Freud, a somewhat comparable process is described as the ego's and superego's growing control over the immediate drives, represented by the so-called id.[22] However, none of these imaginative minds did very much to expose their ideas to empirical scrutiny, although Piaget took some steps to do so.[23]

However, there are a lot of such empirical studies, mainly done by representatives of the cognitivist branch of psychology. Many could be mentioned, but here I confine myself to just one of them: psychologist Walter Mischel. He is well known for his so-called marshmallow experiments, which are empirical studies of self-restraint. In the 1960s Mischel started to test the potential for self-control among preschool children. He has continued to do this intermittently ever since.[24] One of the most intriguing results from these experiments is that children who are able to resist their yearning for instant satisfaction tend to have a better career later in life than those who cannot resist the temptation to have one marshmallow (or other treats) at once instead of two a little later. They also tend to have less economic troubles and better health than children with poor self-control, for instance, manifested in significant differences in body mass index (BMI).[25] These findings have been replicated in other studies, such as one carried out on more than one thousand children in Dunedin, New Zealand.[26]

One might be inclined to trace the different behavioural responses among the children exposed to this test to differences in their personalities, to inborn qualities. But they could just as well be due to different experiences resulting from differences in trust, for instance, those children who thought they had good reason to believe the promise to get two marshmallows later instead of one now were ready to wait whereas those who did not trust the adults to the same degree, perhaps because of bad experiences, were less inclined to wait. They might have displayed a completely rational risk aversion. Mischel was fully aware of this from the beginning, which is why he and his students in the earlier experiments took certain steps to neutralise the difference in trust among the children as an explanatory factor behind their different ability to delay gratification. However, later they also staged some studies in order to be able to substantiate the relevance of trust for the variation in outcome among the children.[27] As Mischel summarised: 'The child's experiences in the first half dozen years of life become root for the ability to regulate impulses, exercise self-restraint.'[28] The implication of this is that the postponing faculty is far from completely prewired. It can be affected and improved among most children on their way to maturation, as stated explicitly by Mischel himself.[29] The mission is to place the so-called cool system (i.e. cognition) in charge of the so-called hot and emotional let-go system, aiming at inhibiting some cravings in the service of long-term goals.[30]

In this context, Mischel's experiments are relevant, not primarily for what they can tell us about the conditions for individual success but for what they contribute to our understanding of the significance of trust and other circumstances surrounding humans' ability and willingness to postpone the satisfaction of their needs and desires. However, if so one must take account of the recent empirical criticism of Mischel, de-emphasising the link between the ability to delay gratification and future success, and showing the significance of social background for the rather weak link that is still discernible.[31] Furthermore, in Mischel's studies a historical view on the matter is missing (no doubt for good reasons), although he applies a highly relevant life-course perspective in his studies. Most interesting here is the close connection between people's imagination of their future self and their willingness to delay rewards: the more continuity people think there is between their current and future self, the less inclined they appear to be to discount future rewards and prefer instant joys.[32] Furthermore, the ability to think beyond the present situation seems to go hand in hand with the ability to think beyond oneself (i.e. to empathise with others).[33] All this brings to the fore questions that deserve being addressed in historical research on the matter, and so far these questions have not been distinctly addressed as *historical* questions.

Due to the disciplinary division of labour in academia, it is no surprise that among humanities scholars, researchers engaged in historical studies are probably those who have devoted most scientific energy to the long-term destiny of self-control. Yet, Norbert Elias, one of the most influential among them, was trained as a sociologist. In the late 1930s, he published his two-volume work on the so-called civilising process in Europe, wherein he studied how the emergence of the centralised state in early modern times conditioned the education of manners in Europe, or in other words, how the monopolised use of violence contributed to pacify the nobility and to eventually make decent citizens out of people, unused to holding back their spontaneous impulses (i.e. unused to delaying the satisfaction of their immediate wants).[34]

Elias had to wait for his breakthrough until the late 1960s, but from that point in time, his studies of the civilising process – it was almost a life-long calling of his – have been a source of inspiration for many historians, sociologists and others. A case in point is the evolutionary-oriented psychologist Steven Pinker, who in his recently published work *The Better Angels of our Nature* utilises Elias's empirical generalisation to explain the long-term decline in lethal violence on both the interpersonal level and the interstate level.

Of course, Elias has also met some criticism, most fiercely from the German anthropologist Hans Peter Duerr, whose major attack, launched in an incredibly long series of books, concerned it being ahistorical to think of people in the past having such little self-control as claimed by Elias.[35] However, as a whole Duerr's attack is a specimen of magnificent overkill. Indeed, by one-sidedly emphasising the disciplining aspect of the civilising process, Elias not only overstated the generality of growing long-term impulse control but also overlooked the fact that the gradual civilising of manners has a two-sided trajectory: on one hand, it moves towards tightening self-control and, on the other hand, it moves towards an increasingly lax attitude to human needs and desires.[36] Moreover, he does not seem to have been aware of the fact that seemingly impulse-driven violence in popular revolts of pre-industrial times was often planned (i.e. prepared for in advance). It was an integrative part of a culture of violence rather than the spontaneous outbreak of rage that it appears to be in retrospect. This is not to deny that once let loose, violence can run completely wild.[37]

Undoubtedly, though, through a multigenerational educational process, members of human culture have learnt to behave less violently towards each other (i.e. to refrain from giving in to their aggressive urges; see appendix B). This firmly established observation corresponds quite well with the predictions implied in Elias's work. A thorough and nuanced test of Elias's theory has recently been carried out by British historian Manuel Eisner, to a large extent reinforcing his results.[38] What Elias did not consider, though, was that the restraining of certain impulses evolved together with the freeing of others. Thus, the pacification of people's interpersonal aggressiveness went hand in hand with the liberalisation of their sexual desires, eventually leading to an acknowledgement of almost all types of sexual preference except paedophilia and incest – not to mention the simultaneous explosion of mass consumption. All of this was based on the emergence of a new set of values considered more humane than the older ones. Instead of acknowledging these intricate trends, superficially seen to be working in opposite directions to each other, Elias tried to explain them away with the aid of a far-fetched ad hoc reasoning.[39] However, nothing of Elias's strained reasoning should overshadow the great value of his contributions to our understanding of the historical trajectory of postponement.

Clearly, Elias is far from being the only researcher who has dealt with matters related to the historical destiny of self-control. For example, in works on the history of mentalities, historians have tried to identify important turning points as to how secular and ecclesiastic authorities have tried to come to grips with people's sinful inclinations, such as avarice, lust, disobedience and so on. The French historian Jacques Le Goff regarded the introduction of purgatory in the European High Middle Ages as such a turning point. In his view, this innovative interlude between Heaven and Hell was an expression of the gradually ongoing individualisation of guilt, of the increasing significance of the intentions behind human misbehaviour whereby gradations from venial to mortal sins were also introduced. Accordingly, the fire of purgatory was described and depicted not only as extremely painful but also as rejuvenating and thus as a potential gateway to Heaven. Thus, in Le Goff's view, purgatory represented a relaxation of society's demands for self-restraint among its subjects.[40]

This interpretation of the medieval zeitgeist is, however, very different from the picture of this period drawn by Jean Delumeau, another French historian of the so-called Dark Ages. In his view, they really deserve being called *dark*, not least because of the authorities' severe attitude towards people's cravings. The church used all means to instigate fear in people so that they would abstain from all sorts of indulgences, many of which we would consider rather innocent today. The primary message was that people experiencing all kinds of calamities in the earthly treadmill could only have an agreeable afterlife through contempt of all worldly matters, through *contemptus mundi* as it was stated.[41]

Now, the contrast between these two historians almost dissolves in view of the fact that Le Goff focused on the relatively prosperous thirteenth century, whereas the *contemptus mundi* discourse addressed by Delumeau mostly concerned the stagnating fourteenth century, in due course heavily struck by the Black Death. This indicates that hardening material circumstances to some extent explain that the authorities' attitudes also grew harder as the promising thirteenth century deteriorated in the increasingly hopeless fourteenth century. Without settling the case, it certainly gives some clues that could lead to an answer to the overarching question of this chapter.

Turning to research on modern times, let me just mention Robert Fogel's findings on the dramatic changes over the last one and a half centuries of working life, as well as on the consumption patterns found in what we now consider the rich world. As people are spending a gradually declining share of their lifetime working for a living, a steadily shrinking minority are occupied in the primary sector. Correspondingly, while a century ago households spent around three-thirds of their income on food, clothing and shelter, today only about 13 per cent is used for such needs.[42] It is very likely that this sea change in human behaviour has had a large impact on people's readiness to delay the fulfilment of their cravings. At the same time, however, this particular matter was not addressed by Fogel nor by any other scholar referring to his work, and since it remains to be demonstrated, it also deserves to be investigated.

Perhaps Fogel's research could be linked to the kind of sociological studies which engaged British sociologist John Goldthorpe and others in the 1960s. Their overarching question was about what happened to the mentality of the 'affluent worker' in the welfare society. What remained of the class consciousness of the working class in such a society and so on?[43] Such questions border as much on the questions asked in this chapter as they do on the ones raised by Fogel, though without providing direct or clear-cut answers to them.

What can be done?

I envision alternative answers to the overarching question raised in this chapter. From one vantage point it is reasonable to assume that people need some belief in a good future in order to be willing to trade instant joys for future rewards. People in affluent and fairly stable societies, having good reasons to predict a prolonged lifespan, might be able and willing to calm down a little, relying on the likelihood that what they do not indulge in today will still be available tomorrow. Accordingly, one might be inclined to predict that when people's expectation of a long lifespan is low, their incentive to suspend short-term fulfilment might be low too.

From another vantage point, however, one might well predict the opposite: increased welfare and a longer life will expand the opportunities for instant gratification, which might reduce the need for the postponement and self-control that are simply necessary for people whose living conditions are not very far above subsistence level.

The question, then, is under what cultural (i.e. political, social, economic) conditions are people most ready and capable of disciplining their inner cravings for instant satisfaction? Do we already have the answer to the question? Against the backdrop of the incomplete account of the state of the art given above, I do not think so.

On the one hand, it goes without saying that psychological and economic research on people's behavioural attitudes concerns the conditions both for facilitating and delaying the fulfilment of their needs and desires, as well as the trade-off between them. However, it is as self-evident that normally such research does not address the fate of self-constraint from a long-term historical perspective, although sometimes it does so ontologically (i.e. over a substantial part of the life-course of people subjected to experimental interventions). Here, Mischel's research is an admirable case in point.

On the other hand, historical research on human needs has not very much focused on the need for abstention and delay. Certainly, there have been quite many discussions of this problem in research on the prehistory of human civilisations, such as notions on the important role played by early priesthoods in limiting the overconsumption of procured resources.[44] Furthermore, as the examples of Le Goff and Delumeau show, there are indeed some historians who have addressed the question. Yet, they are quite few and their time horizon is rather limited, and although the studies of prehistorical resource management apply a far longer time perspective, they do not stretch all the way to modern societies. Furthermore, on the whole, historical research has mainly been preoccupied with how people have managed or failed to acquire much wanted resources rather than with how to restrict access to them.

It should be admitted though that Norbert Elias represents a positive exception in the sense that he applied a holistic approach, as well as a long-term approach, although his studies were not built on a very solid ground and his causal reasoning leaves much to be desired.[45]

So, with my particular question in mind, by joining forces, researchers from different fields devoted to studies of human self-control could make use of each other's strengths as well as overcoming certain field-specific weaknesses. The question could be addressed by a series of internationally comparative and, at the same time, historical studies. That would be a macro-historical approach.

Chapter 5

WHAT EXPLAINS THE WIDESPREAD DIFFUSION OF INEQUALITY AND THE GRADUAL EMERGENCE OF EGALITARIANISM OVER THE CENTURIES?

Introduction

This chapter is about the very long-term destiny of equality and egalitarianism, where equality between women and men is regarded as an integrated aspect.[1] Taking an imagined immortal bird's eye view on the question, one gets sight of two seemingly contrasting processes. On the one hand, there is spontaneously growing inequality, although partly offset by some counteracting trends. On the other hand, there is a gradual long-term expansion and the ever more widespread presence of egalitarian visions of how human society should be organised. Such visions are not being cancelled out by anti-egalitarian world views, which, nonetheless, never seem to become extinct either. To formulate this even more distinctly, one would be inclined to state that an innate need for equality among humans is as evident as a trend towards the opposite in real life.[2]

In his classical treatise *Equality*, originally published in 1931, R. H. Tawney once succinctly expressed this inconsistency:

> Institutions which have died as creeds sometimes continue, nevertheless, to survive as habits. If the cult of inequality as a principle and an ideal has declined with the decline of the aristocratic society of which it was the accompaniment, it is less certain, perhaps, that the loss of its sentimental credentials has so far impaired its practical influence.[3]

How come inequality seems to go hand in hand with egalitarianism? That is the question discussed below that will hopefully be more energetically addressed by the community of researchers than it is today.

The equality–inequality gradient

It is likely that, in the absence of political initiatives, the spontaneous development of society in the long run would lead to continuously widening gaps in the distribution of material resources. In turn, it is suggestive that this would also condition the unequal distribution of other resources (such as health and lifespan) between social classes, as well as between women and men, young and old, and people in sparsely populated areas and people in the urban centres.[4] As recently reported by Augusto López-Claros and Bahiyyih Nakhjavani, one rarely noticed expression of this is that there are more males than females in the world, although the opposite 'should' be the case.[5] This is despite the likelihood that these negative spin-off effects may be partly offset by increasing social equality, such as free access to schooling, subsidised medical care, general suffrage and so on (see below, p. 74). But again, although steadily ongoing, these processes have not yet closed the gender gap in terms of access to social resources.[6]

The assumption behind this way of reasoning is that in a distant past, very small gaps already gave those with a little more than others an advantage in terms of risk tolerance, leading to gradually widening gaps. A stepwise materialisation of the so-called Matthew effect could emerge solely from what was originally an almost infinitesimally small difference in the fertility of the soil between otherwise equally equipped farmers living near subsistence level or at it. This is roughly how Per Molander argued in a study of the anatomy of inequality, adding that both parties have access to different buffers of resources of which, however, they use up the same amount. Self-evidently, this strains the conditions most for the less fortunate.[7]

In a book on war and peace, where the distribution of wealth and income play decisive roles, Peter Turchin argues along rather similar lines. In his model world the 'fatal' process starts with an 'absolutely egalitarian distribution of wealth', after which Turchin asks whether this equality will 'persist into the next generation'.[8] The answer is an outright *no*, and the dynamics he identifies are similar to Molander's. However, in Turchin's model the decisive factor setting the process in motion towards income inequality is the variation in the number of children between families and not, as in Molander's case, the initially small variations in the conditions for crop cultivation. Another minor difference is that Turchin, but not Molander, makes a distinction between hypothetical wealth and income inequality. Whether the former translates into the latter depends on the population density according to Turchin's model.[9] In passing, it could be added that spontaneously growing inequality could be accelerated by deliberate initiatives to curb the 'threatening' trend towards equality, as claimed by Daniel Dorling.[10]

Be this as it may, as pointed out by both Molander and Turchin, whatever the specific drivers behind gradually growing resource inequality, it goes without saying that such a process cannot proceed in a linear way to a state where one individual possesses everything and everyone else has nothing.[11] This is distinct from the model world where resource distribution not only can span the entire distance from being perfectly even to being perfectly uneven, it also has to span it. For example, according to the Gini model, based on the Lorenz diagram (see Figure 2), the coefficient goes from 0, where everyone has as much (or as little) as everyone else, to 1, where the opposite is the case. But both these are theoretical endpoints, and it is as unlikely that the former will ever come true as it is impossible that the latter could ever happen in the real world.[12] The latter does, though, eventually materialise in a simulated world without any redistributive mechanisms.[13]

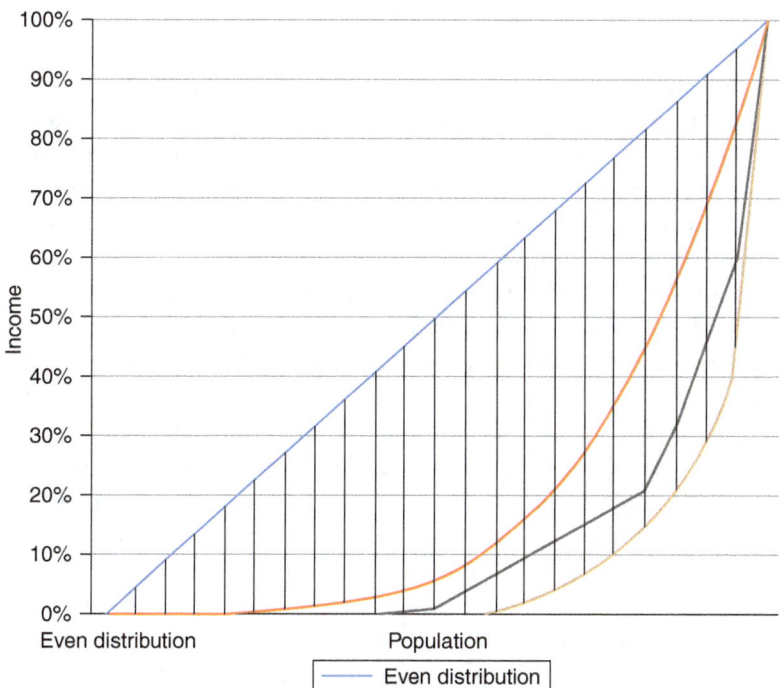

Figure 2. A Lorenz diagram

Thanks to an abundance of research among economists, historians, economic historians, anthropologists and others, today we know a lot about the long-term development of the distribution of resources in the world of real societies. What begins as a spontaneously emerging skewness between small-scale farmers and the like proceeds and accelerates when non-state societies gradually turn into state-based civilisations with expanding bureaucratic apparatuses, standing armies and various elites managing to take hold of growing surpluses from the primary producers.[14]

Cultural anthropologist Christopher Boehm summarises the millennial trajectory of changing resource distribution by stating that 'before twelve thousand years ago, humans basically were egalitarian' whereas for 'more than five millennia now, the human trend has been toward hierarchy rather than equality'.[15] Boehm's basic idea is that humans generally have both a disposition to dominate others and a disposition to dislike being dominated.[16] Among early hunter-gatherers and in tribal societies, this conditioned what he calls a *reverse dominance hierarchy*, where the rank-and-file members of the social group block the way for wannabe alpha male leaders to dominate them too much. Boehm quotes anthropologist Harold Schneider, summarising this built-in ambivalence: 'All men seek to rule, but if they cannot rule they prefer to be equal.'[17] This behavioural pattern is distinct from the pattern discerned among primates where straightforward hierarchies are salient. Boehm's explanation of this difference is that the cognitive capacity of humans by far outsizes and out-competes the brains of primates. In its turn, this provided 'prehistoric humans with a set of intellectual tools by which they could decelerate the trend towards inequality, such as actuarial intelligence (thinking intuitively statistically/stochastically) and displacement (thinking beyond the present)'.[18]

Boehm claims that as equality has been found in all kinds of primeval environments, so inequality has tended to spread all over the world.[19] However, his observations need to be qualified. Indeed, he does so himself, on the one hand by pointing out that archaic equality normally meant male equality in the 'band' or tribe while females and children were in an inferior position within the family.[20] On the other hand, he underlines that some successful attempts to reverse the long-term trend towards increasing inequality have been made.[21] Yet, his overall idea is that of a U-shaped evolutionary curve, from so-called orthodox hierarchies among chimpanzees and the like to egalitarian 'reverse hierarchies' in our species and back again among us.[22] However, Boehm does nothing to validate the latter part of this U-curve, which is fully rational since his major interest is to analyse the mechanisms behind primeval egalitarianism, not modern society. Fortunately, other researchers have done a lot to scrutinise and specify the trajectory of human resource distribution.

In the 1950s American economist Simon Kuznets presented a theory according to which income distribution follows – well, indeed has to follow – a so-called bell curve. It has the same shape as the curve suggested by Boehm unless confined to humankind, leaving the hierarchical apes out, and, most importantly, it is turned upside down in contrast to Boehm's curve. This means that inequality was low in poor and pre-industrial societies and increased in the earlier stages of the process towards industrialisation, only to decrease as the process matured. Kuznets substantiated his theory by reference to an ambitious time series of income distribution in the United States between 1913 and 1948.[23]

Kuznets's contribution to our knowledge of the destiny of income distribution has been very influential. However, today his empirical generalisation is criticised along with the theory behind it. One crucial observation is that of widening income gaps throughout the pre-industrial period, at least in the early modern part of it. This means that the gaps were already wide – and had long been widening – when industrialisation and economic growth took off, which is why inequality and growth do not seem to be causally connected.[24] Therefore, Kuznets's inverted U-curve can no longer be held universally true – or true at all. No wonder, since it covered only one country over a period of 35 years, including two world wars that disrupt the generalisability of the Kuznets curve even more than would have otherwise been the case. So, the 'day might have come to conclude […] that the Kuznets curve is now obsolete'.[25]

The shortcomings of Kuznets's studies are pointed out by French economist Thomas Piketty in his much discussed bestseller *Capital in the Twenty-First Century*,[26] already alluded to above (p. 35). The cause of compressed income gaps during wars, he claims, is not the maturity of industrial capitalism but the destruction of capital in the wake of the wars. This is also reflected in the capital–income ratio, which tends to drop in wartime. Once peace is restored, income gaps begin to re-widen and the capital–income ratio recovers correspondingly.[27] This is generally what has tended to happen spontaneously in the economic history of the world, a notion very much in line with the theoretical reasoning of Molander and Turchin.

Piketty's study covers a far longer period and more countries than Kuznets's time series – in some respects more than 200 years, but mostly 100 years.[28] One of his overarching conclusions is that over the last four decades, income gaps between citizens have widened in a large number of countries.[29] In his book *The Haves and the Have Nots*, Serbian-American economist Branko Milanovic seemed to reach the same conclusion as Piketty: there is no longer strong support for Kuznets's theory when it comes to the post-war period.[30] Accordingly, referring to figures based on the Gini points gathered by François

Bourguignon and Christian Morrion, he concludes that there has been a slight trend towards rising global inequality, not only in recent times but also bicentennially.[31] However, whereas the income gaps are not only widening within and between countries, by being to some extent offset by rapid economic growth in China and India, no certain trend is identified by Milanovic.[32] Here, he seems to reach a somewhat different conclusion to Piketty.

Moreover, in his new book *Global Inequality*, Milanovic argued quite strongly in favour of Kuznets, stating that long-term income data is fairly well compatible with his predictions.[33] First, by tracing data farther back than Piketty, he identified the clearly inverted U-curve similar to that observed by Kuznets. Second, while agreeing with Piketty that the sustained downward trend in inequality largely correlates with the two world wars, he claimed that it could and should be seen as an endogenous part of the economic development itself (i.e. as reactions to a period of rising inequality). Third, Milanovic suggested that the surge in income gaps since the 1980s could also be seen as a second Kuznets wave rather than as a basis for a refutation of Kuznets. It is the combined effect of a second technological revolution and globalisation he claims.

In his earlier book Milanovic also suggested that the real inequality–equality gradient should be assessed not only by income gaps but also by the more or less free access to certain services, such as education. The less accessibility to such services depends on people's income, the less inequality in society, all things being equal.[34] As once expressed by R. H. Tawney: 'The surplus resources of society should be so husbanded and applied that it is a matter of minor significance whether they receive it or not.'[35] Still, Milanovic stated that 'the world is an extremely unequal place where by far the largest chunk of inequalities between individuals is explained by their citizenship'.[36] Phrased by John Roemer, this is an 'exogenous circumstance'.[37]

Doubtless, the information gathered and analysed by Piketty, Milanovic and others about two hundred years of income inequality in a fairly large number of countries has moved the knowledge frontier substantially forward, not least in relation to Kuznets's rather limited but nonetheless highly influential inquiry into its development in the United States over a fairly short period of time. Could the time line be stretched even further back in time? Probably, at least according to Austrian ancient historian Walter Scheidel who, in his recently published book *The Great Leveler*, maps and analyses the trajectory of resource distribution over the past 2,000 years, mainly in Europe but also in the Americas for shorter periods.[38]

Through ambitiously measuring the Gini coefficient of relevant variables within countries or societies over this impressively long time span, Scheidel discerns a general pattern of millennial ups and downs in the distribution of material resources – thus, not only of income distribution per se. His overall conclusion is that stable societal conditions promote inequality, whereas the following four types of unstable circumstances explain the temporary levelling of income disparities: war and mobilisation for war, epidemics, dissolution of the state and revolutionary upheavals. This cycle-like process is illustrated below by one of Scheidel's graphs depicting the very long-run development in Europe (Figure 3).

According to Scheidel, 'over time, inequality increasingly became the norm'.[39] In addition, he states that 'in the apparent absence of mechanisms for the periodic adjustment of renewed inequalities it is perfectly possible that an initially equalized distribution of wealth gradually gave way to more uneven outcomes'.[40] So, as Scheidel's generalising views very much resemble the theoretical predictions of Molander and Turchin, his major empirical observations conform quite well with Piketty's findings about the development of income inequality in twentieth-century Europe and the United States, as well as with Alfani's and Di Tullio's findings about late medieval and early modern Venice (and other parts of Europe).[41] These findings concur perfectly well with current trends in the OECD (Organisation for Economic Co-operation and Development) countries, not least the Nordic countries where, for instance, the income gaps in Sweden have risen even faster than the average.[42]

Against this background, I am inclined to conclude that there is a high degree of consensus among scholars from different disciplines about the long-term development of the equality–inequality gradient. Indeed, when it comes

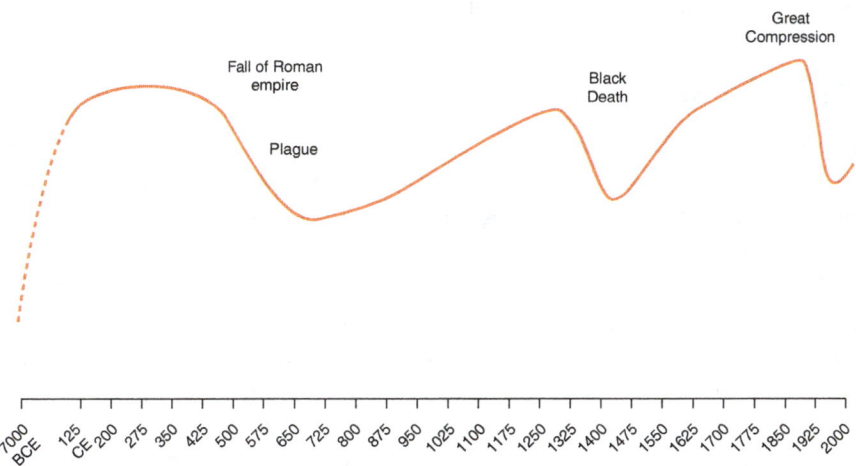

Figure 3. Inequality trends in Europe in the long run

Source: Walter Scheidel, *The Great Leveler: Violence and the History of Inequality from the Stone Age to the Twenty- First Century* (Princeton: Princeton University Press, 2017), fig. 3.1

to explaining the development, there is a lot of disagreement among the researchers. Still, the basic facts are accepted by most of them.

However, perhaps the empirical conclusions reported here should be modified as public in-kind services are taken into account, such as equalised access to social security, subsidised health care, equalised access to schooling, universal suffrage, equality before the law and so on.[43] I agree that such factors should be considered for an overall assessment of what could be called social equality. As noted above, the more the accessibility to such services becomes independent of income and income gaps, the less relevant those gaps become for disparities in living conditions between the better and worse off.[44] However, from such a perspective one should also pay attention to the weakening of such redistributive measures that lie behind the recent exceptional rise in income inequality in the Nordic countries.[45] Furthermore, British geographer Daniel Dorling pointed out that initiatives taken in the early post-war years to weed out poverty were dismantled in favour of regained social exclusion 'when the worst of want was seen to have been eliminated'.[46] Equalised accessibility to certain services was offset by new forms of elitism, such as schooling, which, on the one hand, has been gradually extended until age 18 but, on the other hand, has led to biased access to higher education. Dorling labels this trend 'IQism', supposedly just another word for exaggerated meritocracy.[47]

Now, let us proceed to the second suggested general observation: the age-old presence of egalitarian thinking and egalitarian dreams gradually amplifying and becoming systematised through the times.

The trajectory of egalitarianism

In contrast to the millennial trend towards growing material inequality, when it comes to the long-term change of general attitudes, there seems to have been a change from an explicit affirmation of inequality towards a stress on the benefits of equality between humans.

However, on the one hand, although history from a long-term macro perspective has unfolded this way, a closer look at the process reveals that it has not been straightforward. It is no surprise – trends almost never materialise without minor counter-trends. So, although the idea that all humans are to be considered basically equal has become gradually more widespread and accepted through the ages, the pace at which the diffusion of this idea emerged has varied over time. During certain periods of profound societal development, the acceptance of egalitarian views has weakened.

On the other hand, despite not being systematically thought through, the presence of some egalitarian attitudes is discernible early on, even in the prehistory of humankind, as pointed out by Christopher Boehm and discussed above (p. 70).

Self-evidently, all this means that anti-egalitarian and egalitarian attitudes have coexisted most of the time, constantly challenging each other and differing in relative force. For example, for many centuries – let say from late Antiquity roughly up to the era of mercantilism – the idea of society being like a great chain of being was an overarching ideology. This static *Weltanschauung* implied that people were born unequal and should remain so throughout life and through the generations.[48]

This hierarchical idea is not only primordial and discernible in many parts of the world, it has also prevailed far into modern times, for example in Europe no less than in China. It has often been described as a social system of three fixed orders, such as 'priests, farmers and men of war' (or *ecclesiastica, oeconomica* and *politica*).[49] As Georges Duby said: 'Order is thus the sacralized basis of oppression.'[50] But it was less fixed than one might imagine: from Antiquity until the High Middle Ages it was rather a binary system, simply consisting of the rulers who gave orders – often canonicals[51] – and the ruled who took orders.[52] Gradually it differentiated into a 'trifunctional' system, only to start waning altogether towards the late eighteenth century with the dawn of the Enlightenment and liberal egalitarianism.[53] It took some two hundred years from the dawn of the idea of three fixed orders until its institutionalisation.[54] At the same time, people of all times and places seem to have a strong inclination to divide things in tripartite ways. No wonder then that such system emerged in medieval Europe, as elsewhere.[55]

However, ever since Antiquity many people embraced the Utopian idea of an Arcadia where everyone was everyone else's equal.[56] Moreover, early on, urban journeymen and rural farmers established systems of solidarity that manifested certain aspects of egalitarianism.[57] As a matter of fact, even the gradual establishment of the idea or theory of a trifunctional social system, considered constitutive for a harmoniously working society, contained a tiny rudiment of egalitarianism. Through this mental shift, free but poor peasants were namely transformed from passive and despised recipients of royal protection to key contributors of the means of subsistence for all, the other two orders no less than themselves.[58] A self-evident fact one may think, earlier on curiously overlooked, but in this period about to be acknowledged. This meant that a proto-egalitarian microbe had taken root inside a profoundly hierarchical social body, not to mention the extension of the term *bellatores* so that it denoted not only princes but all warriors.[59] Hence, from a so-called backward-looking perspective, the trifunctional system seems static and rigidly inegalitarian, whereas from a forward-looking perspective it appears as a dynamic system which, little by little, was moving towards less inequality. This meant that the age-old contempt of work and working people were declining – also, little by little. Needless to say, this was nothing as compared to the far-going heretical and extremely chaste egalitarianism popping up and soon getting squeezed in the same period of the High Middle Ages.[60] The trifunctional system was certainly meant to be static and everlasting, although it was not.[61]

However, egalitarian ideas were even present in ancient times. The age-old presence of the Golden Rule is the primeval seed of that. This could be linked to the theory of the so-called Axial Age. It has been claimed by the adherents of this theory that over a period of approximately 400–500 years in the first millennium BCE, certain powerful thought systems emerged in different corners of the world. They were all characterised by a nascent ability among (some intellectually leading) humans to transcend the present state of their societies, as well as themselves and the in-group's self-orientation or self-interest, in favour of considering the needs of ever wider segments of humankind. This transformation of thought could be linked to the independent introduction in many places of the so-called Golden Rule. Today, the insistence of even including future generations in such considerations is commonplace – an extension of that rule.[62] However, the Axial Age theory has never been firmly established and is still to be considered a hypothesis. No doubt such thought systems did emerge in the period in question. It is, though, far from clear that they were widely and broadly adopted by the peoples of this age. Moreover, whatever their degree of dissemination, through the ages, self-reflective, transcendent thoughts have experienced severe backlashes, which is not reflected in the Axial Age theory. Thus, here are a set of questions that deserve to be approached and aim at explaining the ups and down of the Golden Rule, as well as evaluating the validity of the Axial Age theory.

However, the aforementioned examples do not disprove the general observation of a long-term normative change towards an egalitarian mentality. It is manifested in many ways, not least by the fact that modern theories of justice, at least since the eighteenth century and onwards, stress the need to impose restrictions on social institutions that spontaneously generate inequality. It particularly applies to laws regulating the right to inheritance, which some even wanted to abolish altogether.[63]

Also, the general evolution of lawmaking is an indication of such a normative sea change. At an early stage, legal constraints were put not only on the ordinary subjects but also on the power-holders themselves while still within the general framework of inequality before the law. In this way, explicit and outright inegalitarianism was modified, although far from abandoned. Much later, such restrictions were transformed into a universal principle of equality before the law.[64] This was followed up by the introduction of freedom of expression and other elements of human rights. An even more advanced step was taken when human rights were negatively defined, which began to happen in the late eighteenth century.[65] These reforms manifested political equality in the sense that formally all citizens were equally entitled to raise their voice and express any opinion. Furthermore, no one's vote should count more than anyone else's vote. These were substantial steps forward, although human rights have been circumscribed or curbed by nervous authorities over and over again.

As is well known, in due course political egalitarianism was extended into social and economic egalitarianism. It took off with the dawn of the labour movement and industrial capitalism and has continued to play a significant role although being constantly challenged by a more minimalist and formalist liberal egalitarianism.

Another secular trend is a gradual extension of egalitarianism to encompass ever wider circles of people, at certain points trespassing over ethnical and national borders and in due course encompassing all humankind. For example, the 1970s egalitarianism of John Rawls is clearly confined to a fair distribution of resources within nations, explicitly denying its applicability to equality between them.[66] Much the same applies to Ronald Dworkin, although his social-liberal egalitarianism differs from that of Rawls. In contrast, today's philosophers, like Peter Singer, Thomas Pogge and other thinkers, apply a globally universalistic perspective to the problem. As formulated by legal and political scholar Ayelet Shachar, there is no reason why we should accept the 'birthright lottery' that is the rule of today's world.[67] Instead, she envisions a 'universal personhood' that would replace national rights.[68] Of course, such a far-reaching universalism is still an avant-garde position, barely represented by more than a tiny minority of intellectuals, although this minority might be somewhat larger today than in the past. Furthermore, the extension of egalitarianism to a quest for worldwide justice disregarding national borders does not come without frictions. The nationally based egalitarian welfare project of the late nineteenth century and most of the twentieth century is not easily harmonised with the global egalitarianism of our time. Tensions stemming from the goal conflict buried here are heavily enacted in contemporary politics, presently becoming markedly less and less universalistic than earlier in the post-war period. Again, theories in defence of inequality are still marketable, such as the basic preaching of Pareto and Hayek.

Thus, the matter of equality and egalitarianism remain contested issues. Nonetheless, the idea of equality seems to prevail in public and private encounters. Today, even a revival of a Marxist plaidoyer for egalitarianism can be seen.[69] What is the explanation? A general improvement of welfare? This is far from self-evident. Rising welfare and affluence could as well trigger the opposite attitude. Thus, here is a field for substantial knowledge advancement – a clearly identifiable research task. It could be related to other questions discussed in this book but perhaps it comes closest to the question raised in Chapter 4 about the relation between the gradual extension of our lifespan and our readiness to postpone the satisfaction of our needs and desires.

Chapter 6

WHY DO PEOPLE APPROPRIATE AESTHETIC EXPERIENCE (BOTH AS PRODUCERS AND CONSUMERS OF CULTURAL MANIFESTATIONS), AND WHAT ARE THE INDIVIDUAL AND SOCIETAL FUNCTIONS OF SUCH EXPERIENCES?

The issue

Aesthetic actvities are present in all known societies, and it seems obvious that they play an important role as an essential ingredient in people's lives. On the one hand, aesthetic considerations seem to be present in almost all spheres and all parts of life. It is as universally human as they are uniquely human.[1] On the other hand, almost no aesthetic considerations seem to be purely or exclusively aesthetic – they carry a very varied repertoire of other considerations and messages too. For example, to a large extent, group identities are enacted by aesthetic means, such as flags, outfits and so on.[2] Rather often such means are even utilised to inflate essentially petty differences between groups – Sigmund Freud once denoted this the narcissism of small differences.[3] But they are also used as individual means to make a mark, to establish personal distinction, to be famous.[4]

By aesthetic activity I here mean the fine arts and fiction, as well as music and dance and the like, or any combination thereof, but also the aesthetic intentions in any other type of human activity – be they fully conscious and deliberate or not, extraordinary or not.[5] Whatever the genre, the consumption or acquisition of what others have created is as much an aesthetic activity as the creation of such expressions themselves. They are closely linked, and all humans do both, albeit on different scales. Yet, whereas professional art and everyday aesthetics are interlinked, and the aesthetic component is inherent in both, they are not the same. As put by Jessica Lee: 'Art, in the institutional sense, is generally created with the category of art in mind, while domesticity occurs as a matter of course in daily life.'[6] One could add that the primary function of objects meant to be artistic is to evoke aesthetic appreciation among the public, whether evoking pleasant or unpleasant feelings or no special feelings at all[7] – as distinct from Immanuel Kant's notion that the aesthetic experience is equal to what is pleasurable or delightful.[8] The term *public* is inserted to distinguish art from other objects also meant to primarily evoke aesthetic appreciation, for example, children's paintings at the moment when they are shown to their parents. While still not fulfilling all formal requirements for a definition, it works reasonably well.[9] But it also indicates that there is a gradient from art to everyone's aesthetics.[10]

Normally, aesthetic appreciation is the primary function of art, although its creation may be and often is motivated by other intentions, not least to draw people's attention to their 'everyday' lives.[11] As stated by Thomas Leddy: 'Artists are constant and close observers of everyday life.'[12] Furthermore, the arts often have implications beyond aesthetic appreciation – whether intended or not, and whether or not they are beneficial or detrimental to human needs.[13] Conversely, normally the primary function of non-art objects is to meet all kinds of non-aesthetic needs. Their primary function is not to evoke aesthetic appreciation per se although they may have – or, more correctly, most often have – intended aesthetic qualities.[14] As exemplified by Roger Scruton, here referred to by Thomas Leddy: '[A] carpenter cannot say that a doorframe looks right simply because it allows the door to open and it meets health and safety requirements. He also insists that the aesthetic is socially derived and motivated.'[15]

It has even been claimed that beauty is an intrinsic quality of any well-integrated practical solution.[16] In medieval discourse it was even generally stated that there always is a connection between the beautiful and the good – that 'everything that is good is beautiful'.[17] In the former case, a non-aesthetic function must be integrated into the form of the object, whereas this is not necessarily the case with objects of art although this is claimed by many thinkers, not least the medieval masters.[18] For example, mathematical solutions are often especially praised if they are considered beautiful, at least by the guild of mathematicians. But beauty in an incorrect solution is pointless, well indeed, impossible – at least as soon as the error is revealed.[19] This also applies to all kinds of 'utilitarian objects'.[20] On the contrary, works of fine art may be appreciated whether they carry logical consistency or not – in the rare cases where this matters at all.[21] Look at the simple formula below defining how to calculate the number of meetings required given that a specified number of participants are supposed to meet everyone else one time each:

$$N(N\text{-}1) \div 2$$

Then compare it with the painting in Picture 3, where the artist Marie-Louise Ekman visualises an ironic idea of continuous metamorphoses between humans and animals, ultimately from a human female to bird, passing through ape and human male.

I would say that these two symbolic representations both have aesthetic qualities, and we can, indeed, appreciate both. However, whereas the equation would lose its aesthetic attractiveness if it were incorrect, correctness in terms of logical consistency (or in terms of anything) is not very relevant for our appreciation of the aesthetic qualities.[22] It is even the ironic violation of the laws of nature depicted in this painting that makes it so appealing.

Now, the fact that in the professional space contemporary international arts and the intentional aesthetics of everyday utensils and services normally operate in different arenas does not mean that they have always done so. Rather the opposite seems to be the case, as once pointed out by John Dewey, who in

Picture 3. Striptease, by Marie-Louise Ekman

his classical study of art as experience stressed the original unity of almost all parts of life. What has eventually come to be labelled *art* has long since been an activity decoupled from other activities. His work abounds in examples of this. For instance, Dewey points out that in medieval Christian church

> music, painting, sculpture, architecture, drama and romance were handmaidens of religion [...] The arts hardly had a being outside of the church were arts were enacted under conditions that gave them the maximum possible of emotional and imaginative appeal.[23]

One could easily add a huge number of examples demonstrating that people in general care much about the form of the things they equip themselves with or are surrounded by. It applies to non-physical as well as physical representations, for example, to the beauty of a text as much as to the attractive or appalling appearance of a cathedral, a train station or a sports arena.[24] Doubtlessly, people devote a lot of energy and spend a considerable amount of time and money on the form of their objects that is apart from the intended basic function of these objects. Either they are the designers themselves or they choose the objects with the most appealing design created by others. Look at the shoes in Picture 4; they would have been as 'walkable' with green laces as they are with the red ones put on in this case. These red laces are 'a fitting solution to a problem that can be solved in more than one way'.[25] This is where extra-functional aesthetics come in.

Picture 4. Shoes with red laces

Furthermore, as illustrated by the shoes in Picture 4, it is obvious that people are not only anxious about the appearance of the things that they think could be associated with themselves as an outflow of their choices – aesthetic or not; in general, from early childhood until very high age, they care as much about how they themselves appear as they do about the aesthetic appearance of the things and people around them. Again, it may concern semi- or non-physical extensions of people's choices as much as physical self-representations: one's verbal style – written or spoken – as well as one's choice of vehicle, outfit (of course), skiing style, gestures, smell, how the dinner table is set and so on.[26] As stated by Yukiko Saito: 'Often guided by aesthetic concerns and interests, we engage with objects around us by cleaning, organizing, mending, rearranging, relocating, and eating on a daily basis.'[27] She also stresses what she calls 'appreciation directed toward ambience', such as the tactile sensations and visual impressions of everything present at a dinner table.[28]

The row of examples is potentially never ending. For example, to furnish a home according to one's taste and to arrange the fairy tale one tells in a certain way in order to make it attractive to the interlocutors, listeners or audience should be considered aesthetic activities no less than the professional dancer's performance on stage.

Now, one might think that the aesthetics of the human mind, seemingly at work almost from cradle to grave, is something peculiarly modern, evolving as growing prosperity allowed people the luxury of indulging in beauty and not only confining themselves to optimising the function of things. But this is very far from the truth. As demonstrated by archaeologists, aesthetic considerations can be discerned very early in human prehistory. One paradigmatic example of this is the introduction of the so-called Acheulian cutting tools about 1.5–1.7 million years ago. Replacing the Oldowan tools, these hand axes not only represented a considerable step forward in the technical sense but also bore the first signs of deliberate design. This indicates at least some rudiments of deliberate aesthetic intentions beyond purely functional considerations.[29] Then, as human culture evolves, aesthetic design gradually becomes an intrinsic aspect not only of almost everything human-made but of everything humans relate to, human-made or not. It applies to wild nature as well as to nature domesticated by humans.[30]

The essential questions

Why do people in general care about how things appear, that is, about whether things are good or bad looking?[31] This is the first question to be addressed: whether such concerns are about people's appearance or the appearance of others, whether they relate to inanimate things or living creatures and whether such considerations are enacted as something called art or not. In addition, it is as interesting with aesthetic experiences that are 'frameless' as it is with experiences directed towards 'clearly demarcated objects'.[32] In this context, *to care* should be understood in a broad sense: not only as people's aesthetic views or attitudes but also as all kinds of action taken to design the world in accordance with the aesthetic preferences. Second, what are the aesthetic considerations about? Perhaps, the criteria once pointed out by art theorist Roger Fry could be applied here: rhythm, mass, space, light and colour.[33] Third, and as importantly, what is the essential function of everyone's everyday aesthetics, and what is its outcome and impact? Fourth, since much action taken to fulfil human needs often has unintended consequences, I would welcome studies that relate the first question to the second one. Thus, is there a connection between the aesthetic need and the outcome of aesthetic action? If so, what is the nature of the connection? If the connection is weak or non-existent, why then is such agency sustained?

Under the umbrella of these overarching issues, one could add questions such as the following:

- To what extent are people ready to sacrifice function for the sake of beauty or other aesthetic values and vice versa?
- Which are the situations wherein it is seen as either legitimate or illegitimate to devote energy to aesthetic value?
- Do certain cultures and historical periods encourage aesthetic investment in social life while other cultures discourage it?

By asking such questions, I want to make it clear that it is the ever-present aesthetic urge or inclination among all humans that is my overarching interest. Implied here is a notion that art is just a subcategory of this almost universal urge or inclination. As expressed by Thomas Leddy: 'Everyday aesthetics is a *branch* of aesthetics.'[34] Like some of the inside scholars in the field, I think it would be a good idea to dissociate the former from the latter. By such a step it becomes possible to avoid the fallacy of focusing on attempts to identify what is artistic or non-artistic in everyday aesthetics. In regard to my specific knowledge interest here, it is a non-issue. From my perspective it would be a waste of time to invest intellectual energy in such conceptual fine-tuning. Yet, subsuming art as just one branch in the tree of aesthetics does not mean that art is uninteresting – not at all. It only means that it considered *a subcategory*.

The state of the art

It goes without saying that the questions suggested above have been raised many times before. Art historians, musicologists, literary scholars and others have provided important knowledge on the aesthetic endeavours of humans. Furthermore, intellectual history abounds with philosophers and other thinkers who have presented a rich repertoire of ideas and theories on art and aesthetics.[35] Still, I think that the questions have been insufficiently addressed.

First, despite all attempts through the centuries to develop a general theoretic understanding of human aesthetics (including taste and art), theoretical work on the question is still insufficiently developed. There are some works on specific modes of aesthetic expression, such as palaeontologist Steven Mithen's speculative ideas about the evolutionary significance of music as well as psychologist Steven Pinker's claim of its evolutionary insignificance. Another example is literary scholar Lisa Zunshine's theoretical idea on humans' propensity to read fiction as a manifestation of our unique capacity for effortless mind-reading of others (= theory of mind).[36] Perhaps John Dewey is the modern thinker who has come closest to an encompassing and coherent theory – a theory ultimately devoted to art and not to aesthetics in general. To him art fulfils a universal human need to regain emotional harmony or equilibrium in a dynamic world where it has got lost and been regained over and over again. The need is fulfilled by both doing and undergoing what he called *an experience* that has a definite end and not only a beginning. Art is expressive; it is the medium through which experience is communicated from the artist to those appropriating it, and its specific form brings it to fulfilment.[37]

Second, however, while a theory or the fragments or rudiments of an overarching theory have been presented, it has not been done with the aid of rigorous tools aiming at empirical operationalisation. It is true that vivid discussions on these matters have run in many directions, year after year, century after century. As far as I know though, very few empirical investigations have been made, not to mention relevant empirical inquiries.[38] Even very interesting and potentially promising attempts at theorising stop short of real investigations. For example, Roger Scruton's idea of beauty – having a fourfold object comprising desire, contemplation, everyday beauty and taste – is well argued and provides suggestive illustrations but is not followed by systematic empirical inquiries.[39]

Third, until recently aesthetic scholars have mostly been preoccupied with high culture, such as novels, paintings and so on.[40] A paradigmatic example of this is John Dewey's classical study of art as an experience. Undeniably, one of his overarching missions was to show that all aesthetics are anchored in people's ordinary experiences. Aesthetics, he stated, are 'the development of traits that belong to every normally complete experience'.[41] For good reason he has even been called 'the grandfather of everyday aesthetics'.[42] Nevertheless, by tracing fine art back to its true origins in all kinds of normal human dealings, it was precisely the emergence and development of fine art that Dewey wanted to explain. This was his major mission, distinctly asserted in the first chapter of his book *Art as Experience*: 'In order to *understand* the esthetic in its ultimate and approved forms, one must begin with it in the raw.'[43]

Still, it cannot be denied that Dewey through his seminal work eventually opened up a whole new field of research on what today is called *everyday aesthetics*.[44] He is the one to be credited for the initiative. The field itself began to emerge some twenty years ago, and today it engages a considerable number of scholars. One of them, Kevine Melchionne, points out the importance of the field in the following way:

> In relation to what is offered by great works of art, the satisfactions of everyday life may be modest. However, their pervasiveness in our lives makes them important, arguably more important than what we gather from conventional works of art that we encounter from time to time.[45]

In my view, this is a promising statement. Unfortunately, however, instead of propelling empirical research on the matter, the 'everyday turn' of aesthetics has triggered endless debates on what should be meant by *everyday*, *aesthetics* and a lot of other related concepts.[46] To Melchionne, one of the most influential researchers in this field, the distinction is crucial.[47] To others it is not.

This self-perpetuating dissent among the aesthetic scholars of everyday life is regrettable. In fact, I find it sterile and tiresome and largely a waste of time. This is so, first because the near obsession with the meaning of *everyday* tends to divert the interest from the universal presence of aesthetic action, whether it happens every day or not. In my view it is not necessary for the researchers to confine themselves to aesthetic endeavours that take place each and every day (i.e. aesthetics as an intrinsic aspect of the trivial or non-special parts of our lives). The important thing is that such action is taken *by everyone*, by people in general. Aesthetic intentions and actions appear among everyone, whether every day or only some days, whether on special occasions – where aesthetic considerations are explicit and conscious – or as repeated semi-conscious routine measures.[48] This does not mean that an aesthetic experience needs to be extraordinary in order to be categorised as an aesthetic experience. One should rather simplify the entire issue: when someone pays specific attention to her or his experience, it is *an experience*, whether extraordinary or not, and it does not necessarily need to be 'rounded out', as claimed by John Dewey, in order to qualify as an aesthetic experience.[49] The key thing is why the human mind is generally and continuously preoccupied with aesthetics and to what extent there is a trade-off between function and form.[50]

Second, most of all the articles and even whole monographs devoted to this never-ending discourse on definitions and delimitations are as empirically empty as the scholasticism of medieval Europe, which it, indeed, resembles quite much.[51] It drains intellectual energy that otherwise could have been spent on real investigations of real phenomena. Certainly, in the field of fine arts there are a lot of empirical studies which are of less interest to me here. In the field of everyday aesthetics, however, solid empirical research is still missing. I fear that in the relative absence of empirical correctives disciplining the discussion, the lofty, opinionated exercise will just go on and on. This is a pity since everyone's aesthetic considerations and actions have a great impact on society and its destiny. Ordinary people's ordinary aesthetic concerns may even be more pivotal for their identities, views of the world and how they pursue their interests and passions in the world, than their experiences of high art.[52]

Ideas

Now, what is the background of the ever-present aesthetic cravings of the human mind and human practice? One might be inclined to explain it in evolutionary terms as a multifaceted repertoire in the service of sexual selection.[53] If so, beautiful people who provide themselves with beautiful things would find it easier than others to find a partner. Sexual selection may play a certain role here, although there are a lot of counter-instances to this simple Darwinian idea. Obviously, a human's outfit plays a similar role in dating and mating as that of the too often repeated example of the unpractical but sexually appealing tail of the peacock. On one hand, however, there are a lot of aesthetic activities which do not have even a remote connection to pair bonding or procreation. On the other hand, as stated by Roger Scruton, 'activities short of artistic creation would make an equal contribution to such a genetic strategy'.[54]

A less evolutionary idea would be that everyday aesthetic activity may be perceived as the human individual's way of transcending the immediate circumstances in which she or he is immersed.[55] This implies a uniquely human ability to imagine what does not yet exist (and might never exist), and maybe it is precisely this ability and this apparently irresistible craving that explains the great significance of aesthetics in human life. This striving to go beyond what is at hand also breeds the creation and appropriation of new forms of aesthetic expressions.[56] The countless opportunities for new forms of expression that the modern computer enables is just one example of that. Appealing as it might be, humans' use of their imaginative potential is not limited to aesthetic activities. It is more general than that.

A related idea is that aesthetic endeavours may bring a feeling of order and lasting stability to a seemingly messy and incessantly changing world, crowded with beings fully aware of their mortality. By trying to freeze the ongoing predetermined decay of life through the composition of seemingly everlasting things or ritualised immaterial cultural expressions that are meant to last, people might manage to attenuate their existential concerns, their death angst.[57] However, rather often aesthetics – everyday aesthetics no less than fine arts aesthetics – serve the opposite purpose: to instil resignation in the human mind, both in face of the recurring disorder of life and of its predetermined termination. Paintings in the tradition of *nature morte* are the most pregnant expression of that.

No doubt other such needs and explanations could be added. At this point I can neither tell how to settle the hypotheses mentioned above nor what other theories would suggest. But whatever the nature of such theories, they will soon become irrelevant and forgotten if not exposed to empirical scrutiny or empirically refuted.

III. WHAT LIES AHEAD?

Thematic clusters

It is rather easy to discern certain thematic clusters among my suggestions (including the ones in appendix B).

First, I have repeatedly addressed matters of societal transitions, whether enacted as dramatic leaps or as gradual changes. This testifies to an interest in identifying processes that do not evolve cumulatively but appear as intermittently recurring constancies throughout history.

Second, the above observation implies a preoccupation with the dialectical nature of cultural evolution, partly enacted as setbacks and even collapses that for some time suspend a process directed towards progress in a certain sense of the word. For example, during certain periods of history the secular trend towards declining interpersonal violence, growing acceptance of 'the other' and technical progress are curbed and even turned into the opposite trend. These turning points need to be explained.

A third interest concerns the mechanisms behind minor norm changes as well as major norm changes. Fourth, many of the questions revolve around the evolution of human needs and desires, both unmediated and mediated.

Last but not least, issues related to the societal significance of knowledge, knowledge advancement and resistance to knowledge should be mentioned here, being a kind of meta question (however, only explicitly addressed in one of the topics discussed here).

Such clustering gives an indication of the particular research interests that have guided me in the choice of questions. This is how it should be. It reveals and invites a critical examination of my scientific idiosyncrasies. Perhaps it also inspires some researchers to suggest that certain questions should be added, while others should be either modified or even removed from the list altogether. I welcome such initiatives. I would even appreciate it if exercises such as mine were carried out in other parts of the world, enabling academics worldwide to get sight of their own as well as other researcher's culturally biased scientific preferences.

Hopefully, repeated encounters around these questions would contribute to triggering a thought process which I think would be beneficial to science and society. Needless to say, I do not intend to exhort all researchers all over the academic world to gather around a limited set of questions, such as the ones suggested here – no more than Hilbert wanted his fellow mathematicians to exclusively focus on his 23 or 10 favourite mysteries. But I do think that the scientific community, like human society at large, would have much to gain if we took on the responsibility to focus more than hitherto on partly neglected or insufficiently addressed, and at the same time profoundly important, questions. This is so since the major research questions discussed here largely correspond with the most vital issues of humankind.

As declared at the outset, my mission has been to formulate questions and not to provide answers, which I, nevertheless, hope that other researchers will try and do. Answers are as important as questions – well, in fact, they are more important.

As hinted at above, certain obviously significant questions have been omitted from this little treatise. Some humanities scholars may be wondering why. Before concluding with a discussion of what humanities are and should be, and what they are not, I will briefly touch upon the reasons behind this decision.

The omitted questions

Apart from questions that are not addressed because of my ignorance, there are basically two reasons why important questions have been left out.

The first reason is that some profound questions have already been addressed and much treated in humanities research. This applies to the most profound question among questions, that of cultural evolution,[1] an almost uniquely human phenomenon. It has been addressed many times over the years, and today it is a flourishing field of its own wherein humanities and natural scientists make concerted efforts to understand the drives behind cultural evolution as well as its consequences. Other such themes are certain linguistic topics, environmental humanities, gender research and so on. Since this is the case, these topics or themes fail to meet one of the criteria presented in the introductory chapter (p. 11). There is no need to urge the community of humanities scholars to do research on questions that are already on the agenda.

One such question concerns what explains large-scale migration flows over the very long term, about which I originally planned to include a chapter. It was motivated in the following way. With the aid of a whole set of new methods and techniques, like DNA analysis and the use of mass data and very powerful computers or grids, it has become possible to pinpoint with great accuracy how people have wandered and migrated in the past, which groups of people have immigrated to different places and which groups have become assimilated or gone extinct/disappeared. The well-known historical fluctuations in migration flows are due to the varying pace at which societal changes take place. Here I envisage explanations, still lacking, of the long-term variations in the range and intensity of those large- and small-scale flows of people. A large subset of questions could be helpful in getting closer to such explanations: Why do certain groups become permanently settled in certain places whereas others move out? Are there certain niches that people fill that have certain comparative advantages? To what extent are the flows of people a matter of expulsion by other people? It has been frequently claimed that the groups with the highest population density will wipe out those groups with a social structure that conditions lower population density. This, then, should mean that hunters would be replaced by sedentary farmers, which, however, is far from always the case. As a matter of fact, during certain periods, nomads with few people per moving square kilometre have expelled groups with a higher density. This is just one example of a question aiming to understand the causes of long-term migration flows.

Not all of the aforementioned questions on migration flows have been addressed, but reasonably many have – enough to make it unwarranted to remind humanities researchers of the significance of these questions for the understanding of the human condition. So, the topic was dropped.

Second, a number of questions have been left out not because they do not deserve being tackled but because they were originally meant to be handled by my intended co-author who unfortunately eventually had to leave the project. However, they are included in appendix B.

What the humanities are and what they are not

Anyone interested in the human condition should go ahead and inquire into it, whether a humanities scholar or not. Ideally, a common set of questions should be addressed by concerted efforts carried out by researchers from different disciplines – given that their efforts are relevant and important for the advancement of answers to the questions at stake. Here it must be stressed, though, that an interdisciplinary approach should never be a purpose of its own. Focus should always be on the question, whether it leads into such efforts or not. But given that it does, such collaborations may inspire and force humanities scholars to reach out of their own terminological monastery, making themselves intelligible also to outsiders.

However, we cannot expect this to come true very often yet. It is not likely that many chemists or physicists will enter the field today. Thus, in the foreseeable future, studies of the destiny of humankind will remain the scientific playground mainly for researchers primarily trained in the humanities and secondarily in the social sciences. Although I am hoping that a change will happen sometime, the current situation is nothing to complain about – it is business as usual and it must be coped with for the time being. My major complaints concern something else, namely the distorted understanding too many humanities scholars have of their specific role in the academic community. I will finish my analytic survey of the big questions about the human condition with a very brief discussion of some proliferate misconceptions of what research in the humanities is and is not, or what it should be and should not be.

Disturbingly often, my humanist colleagues declare that their *special* mission is and should be

- to be *the* advocates of humanist values
- to shoulder the responsibility of being *the* critics of society
- to uphold a belief in human free will
- to integrate humanistic research with performative culture
- to ask questions *instead* of providing answers
- to reflect on human life *instead* of making findings
- to stress that our results are provisional
- to point out the (unfathomable) complexity of human life
- to make interpretations rather than establish facts
- to look for the uniqueness of human life *instead* of its general features
- to apply qualitative methods rather than quantitative methods.

Moreover, many of the notions listed above appear together as a reasonably coherent set of ideas. I find all of them problematic one way or another, together as much as one by one, whether coherent or not.

First, the special mission for humanists *as researchers* is not to advocate humanist values. Instead, it is to provide knowledge of a certain kind. At the same time, I hope that most humanities scholars, nonetheless, nurture humanistic values, and I assume that they are willing to speak up for them. But this should be their mission *as citizens*, no more and no less than it is every citizen's. Thus, I expect the physicist to be as humanistic as the philosopher or the literary scholar, and it has nothing to do with their specific scientific expertise per se. So, to be humane is no prerequisite for being a humanities scholar. An immoral humanist may be as excellent a researcher as a depraved chemist. It needs to be added, though, that this is wholly compatible with the basic motivation for the entire academic business to satisfy our extra-scientific human need to gain knowledge about the phenomenal world, be it animate or inanimate. Ultimately, all scientific curiosity is anchored outside science. This is so, since the choice of one research subject instead of another among an endless number of unknown but knowable phenomena cannot be the lack of knowledge itself.

Second, as a corollary to the first point, I do not think that scholars of our craft has any special responsibility to act as critics of society. It is just a special case of the first point. At the same time, it needs to be underlined that such ideological temperance should not prevent us from pointing out or predicting the knowledge-based consequences of political decisions, nor from voicing warnings about the potentially devastating effects of such decisions in relation to their intended outcomes. Still, our basic role is to offer knowledge to the citizens that they may apply at their discretion – be they policymakers, researchers or something else.

Third, humanist researchers are not by definition advocates of human free will whereas, for example, biologists should be predetermined to represent the opposite view. Yet, such simplistic distinctions pop up every now and then.[2] Sometimes it is claimed that humanities scholars uphold the dignity of humanity while natural scientists reduce us to pure victims of blind natural laws. But such distinctions do not hold. On the one hand, far from all humanities scholars subscribe to the presence of free will whereas, on the other hand, far from all scientists deny it (if they take it into consideration at all). So, the defence of free will cannot be used as a distinctive feature of the humanities. Personally, I believe that only an unexplained will can be free.

Fourth, over the past three to four decades, artistic research has emerged and become established as a recognised academic field, eligible for funding. As far as I understand artistic research, it is a kind of hybrid between performance and the study of performance. My impression is that its practitioners and spokespersons are rather undecided as to whether they want to be an integrated part of the scientific community or not. On the one hand, they organise their activities very much along conventional academic lines, for instance, through postgraduate training and the production of doctoral theses that are defended at public events. On the other hand, in order to avoid being too 'scientificated', some representatives of artistic research state that they do not do research *about* art but *in* art. I am not sure what is meant by this rather mysterious distinction. If it just means that they are studying things while enacting them, it is trivial, but if the distinction aims at dispensing them from presenting testable and thus refutable observations, it is completely unacceptable. Humanities researchers must not give up the clear demarcation line against such an anti-scientific attitude, and they should never get involved with such so-called research.

Without doubt, scientific studies of aesthetic experiences and all sorts of artist performances are major thematic areas for the humanities (see

Chapter 6). At the same time, research on these issues should be kept distinct from these experiences. Like the oceanographer, we need a certain distance from our object of study to get sight of its overall structure, which is and should remain our essential mission as scientific intellectuals. For this sake, it is often (if not always) better to stay detached rather than be too attached. To read and enjoy (or get annoyed by) novels, plays, movies and so on is something different from making systematic analyses of them. Yet, these two easily distinguishable activities are astonishingly often confused when it comes to the identification of what the humanities are. For example, some years ago I listened to a talk by Eric Kandel, Nobel Prize laureate in medicine in the year 2000. The topic was the relation between fine arts, such as painting, and psychiatry. Kandel stressed the need for linkages between science and the humanities. However, to him *the humanities* did not mean research on art, but art itself. This happens over and over; for instance, when the humanities were introduced to the curriculum for medical students, it just included a couple of fictional stories to be read alongside the normal textbooks. Representatives of the humanities should not contribute to this confusion between the consumption of culture and research of culture, and we should shun getting connected to the kind of artistic research that does not acknowledge even the most rudimentary and basic rules of science.

Fifth, if ours is a serious business, basically motivated by extra-scientific human needs to gain knowledge about the world inside us as well as outside us, then the notion that it is more important to ask questions than to answer them is a luxury that we cannot afford. For example, is it more important to ask why racism is so widely spread than to get an answer to the question? Or is it better not to gain reliable and useful answers to the question of how humankind might slow down devastating climate change than to gain them? Of course, the answers to good questions often trigger new questions, but to think away the quest for answers would mean that curiosity evaporates too.

Sixth, some of the interviewees in the *Humanities World Report* stated that a humanities researcher's primary task is to make reflections on the human condition rather than to have findings. Even those who acknowledge that we sometimes manage to give fairly trustworthy answers to our research questions after all – which in my conceptual world is equal to findings being made – frequently stress that such answers can never be more than provisional, as distinct from what is pretended to be the case in the natural sciences. A frequently held position, closely connected to this, is that the major approach in the humanities is to make interpretations, not to establish facts.

These notions both misrepresent what goes on in the natural sciences and overstate the uniqueness of the humanities.[3] As findings are made in all scientific fields, these findings are everywhere based on interpretations – the hypothesis about the Higgs boson no less than hypotheses of what motivates human behaviour. To make an interpretation is simply to infer from what is obvious that which is not obvious. Imagine any researcher in any field who does not do that. This is the air we are all breathing each and every day. This is basically what reflections are all about, whether they are made in order to formulate hypotheses and to make speculations about the facts a priori or to go beyond the established facts through reasoning a posteriori.

This means that, on the one hand, reflections about the phenomenal world and, on the other hand, discoveries of factual circumstances in that world are both highly interdependent and not mutually exclusive activities. It is a matter of a self-perpetuating and never-ending feedback loop. It is almost embarrassing to point out such a self-evident fact. Finally, anyone with some familiarity with the history of science knows that no scientists conceive of their results as definite truths, although some findings appear definite in view of their long-term sustainability. For instance, it seems unlikely that the earth will ever regain its status as being the centre of the universe. Thus, the fallibilist attitude displayed among humanities scholars is something they have in common with most scholars elsewhere in academia. But it is certainly not a relativist attitude: the earth cannot be both round and flat, the centre and not the centre of the universe; witches cannot be existing and non-existing at the same time; the corona virus simultaneously contagious and not contagious, and so on.[4]

Seventh, many humanities scholars state – correctly in my view – that their research matters because they deal with what it means to be human, like the social scientists but unlike the natural scientists. However, by overstating the differences and complexity of human life in comparison to other phenomena, many of these humanities researchers shun the generalising implications of this statement. It is as if our overarching task should be to point out the rich variation between cultures, without ever trying to discern what humans have in common across cultures and cultural differences.[5] Accordingly, this is the reason why so many humanities researchers unfortunately avoid applying statistical methods in their work.

Doubtlessly, human social life is as restlessly dynamic as it is incredibly varied. It is immensely complex in the sense that it is characterised by coordinated large-scale interaction between functions and people.[6] But this is nothing unique; it is rather typical in the phenomenal world. The climate system is possibly as complex and varied as human life. This has, though, not prevented climatologists from trying to look through what seems to be a mess in search of its general patterns. To look for general patterns is to simplify, which is what researchers should see as one of their overarching missions.

Biotic evolution is brought forward by natural selection, which is a simple mechanism precisely based on variation. Without variation, there would be no evolution (as there would not be evolution without heredity either). By the ingenious minds of Darwin and Wallace, the observation of variation in nature became integrated to a general evolutionary theory, instead of being used as a pretext for the kind of particularism that characterises too many humanities scholars. So, the notion of variation should not be regarded as the negation of generalisation.

Now, nothing of what has been said above on what it should mean (and should not mean) to be a researcher in the humanities should be understood as if there are no differences between different fields of research. However, the dividing lines cannot always be drawn between the humanities and the natural sciences. I will conclude with a very short note on this.

To study the human condition, historically as in the present, means to inquire into the feedback loops of the circumstances conditioning and conditioned by unique decision-making, meaning-seeking and culture-building agents. It is thus the subject matter that distinguishes the humanities from other fields of research. Thus, it is not the methods. For example, CO_2 analyses in archaeology have not more in common with textual analysis in literary studies than with methods applied outside its realm, although both are disciplines within it. I would even claim, that it is detrimental to the quality of the humanities to mark out certain methods as *the* distinction, such as the statement that humanities scholars mainly apply qualitative methods, whereas scientists are quantitatively oriented. Apart from being incorrect, such a creed runs the risk of cutting them off from a lot of useful tools to be applied where needed.[7]

So, there is no common methodological denominator for the humanities, beyond the fact that most researchers in the field use the human individual as their analytical base unit. Together with its specific subject matter, this set the humanities apart from other sciences, such as neurology, also focusing on humans. It does, however, not apply to all either, for example, not to philosophers doing semantics, historians studying the interplay between microbes and humans, or psychologists, phoneticians and others doing brain imaging.

Furthermore, humanities cannot be distinguished according to a certain and distinctive epistemology either, which on the contrary should basically be the same everywhere. All should comply with basic rules of logic, and all disciplines devoted to empirical research must live up to general standards of reliability and validity.

Methods differ because the preconditions for inquiry differ, but the differences in such preconditions do not follow the dividing lines between faculties. Historians, palaeontologists, astronomers and others devoted to the study of the past have only indirect access to their objects of study. On the contrary, green biologists and anthropologists gather knowledge about the processes to which they are direct eye-witnesses. These processes would take place even in the absence of the curious researcher, although they may be and often are affected by their presence. This is not the case with experimental studies in physics, psychology and elsewhere, where certain processes can be arranged in a controlled way in advance and then observed in real time. Obviously, the varying conditions for research, as pinpointed here, require different methodological approaches (not to be addressed here), which, again, do not correspond with the conventional divisions of faculties.

For the development and preservation of an open-minded and knowledge-affirming climate in academia, it is important to raise the awareness of these cross-cutting distinctions.

Finally, it is important to pay attention not only to distinctions but also to common grounds between faculties and disciplines. In the academic community there is today an exceptional chance for a rapprochement between the human, natural and technical sciences. This could be built on the widespread recognition of the plasticity of the human brain, long since assumed to exist by humanities researchers and now empirically established in the cognitive and neurosciences. Most of what we humans know is learnt through our experiences, and less and less is considered hardwired from birth. This could be a common ground for the acceptance by humanities scholars of what superficially seen appear as the opposite side of the same coin: that some of our traits have a genetic ground. That so much is acquired in human life and culture is explained by the fact that our plasticity is not. The unique human ability to produce culture and cultural change without having any modification of our genome as a precondition is itself genetically conditioned. Here is a basis for mutual respect and understanding between different fields of research.

APPENDICES

A

FIVE THEMATIC CLUSTERS SUMMARISING A WORKSHOP ON BIG QUESTIONS[1]

1. What makes society possible or impossible?
2. The implications and lack of implications of knowledge
3. Creativity and societal change
4. The significance of aesthetic experience
5. The relationship between nature and culture, and natural and cultural selection

B

TRANSLATED HIGHLIGHTS FROM AN ARTICLE ON THE BIG RESEARCH QUESTIONS

Note: The following was jointly written by Janken Myrdal and me.[1]
[–]

1. What explains large-scale migration flows over the very long term?
[–]

2. What explains that some kinds of knowledge are widely accepted whereas other kinds of knowledge are rejected?

The overarching aim is to understand and systematise the conditions for the spread and acceptance of knowledge, as well as the rejection of knowledge – the notion of *knowledge* here taken in a broad sense. Is it possible to discern dissimilar or even contrasting intrinsic traits of knowledge that generally either invite its adoption or trigger its repulsion per se (i.e. irrespective of its specific cultural or historical context)? Self-evidently, counter-intuitive knowledge is more difficult to assimilate than knowledge that suits people's preconceptions or ideological leanings. But could other such traits of knowledge be identified that affect its varying reception or impact? In addition, what is the significance in this respect of certain pivotal situations, such as certain societal atmospheres, certain human experiences or attitudes? Finally, which are the general cognitive, emotional and ideological factors that can help explain the adoption of knowledge, as well as the repulsion of knowledge?

One relevant circumstance is that the growth of the total stock of knowledge seems to be far more rapid than the growth of what an individual human being is able to incorporate. This means that the gap inexorably widens between the former and the latter. This might speed up the incessant process of specialisation, its wake more than ever forcing people to select what knowledge to adopt and what to neglect. This might reduce or block people's readiness to encompass an increasingly vast mass of information, even risking alienating people from an interest in gathering knowledge altogether.

3. Why do social norms change despite the fact that their mission is to be sustained?

Generally, human interaction is regulated – if not from the very outset in the development of all societies, at some early stage. Such regulation is enacted both by means of physical power and with the aid of norms. But to a substantial degree, physical power also relies on norms since it would not function without the loyalty of strategic groups and ultimately loyalty from the subjects or citizens in general. Norms serve to stabilise social interaction, be it peaceful or characterised by more or less violent conflicts. If such stabilising measures are successful, they will strengthen the norms through a series feedback loops. Norms exist in order to be preserved through vertical and horizontal transmissions and thus promote human action in accordance with the norms. But they change, nevertheless, some more than others and more so in certain situations and under certain circumstances – even those that appear especially stable, sometimes even timeless. For example, in the very long run it has become less and less acceptable to expose children to violence from adults. This has also meant that early in the lives of the children, parents and other adults have to negotiate with them, often by resorting to rational arguments. This may have contributed to the generally improved sequential thinking among children and young people, indicated by the so-called Flynn effect among other things.

How are the dynamics of normative change to be explained? And why are some norms more stable than others? These are the questions to be asked here.

4. Why do people appropriate aesthetic experience (both as producers and consumers of cultural manifestations), and what are the individual functions of such experiences, as well as the societal functions?

Aesthetic activities are present in all known societies, and it seems obvious that they play an important role as an essential ingredient in people's lives.

By *aesthetic activity* here we mean fine arts and fiction, as well as music and dance and the like, or any combination of them, but also aesthetic intentions in any other type of activity. Whatever the genre, consumption or acquisition of what others have created is as much an aesthetic activity as the creation of such expressions themselves. They are closely linked, and all humans do both, if on different scales. For example, to furnish a home according to one's taste and to arrange the fairy tale one tells in a certain way that makes it attractive to the interlocutors/listeners/audience should be considered aesthetic activities no less than the professional dancer's performance on stage.

In that particular sense, such everyday aesthetic activity may be seen as the human individual's way to transcend the immediate circumstances in which she is immersed. This implies a uniquely human ability to imagine what does not yet exist (and might never exist), and maybe it is precisely this ability and this apparently irresistible craving that explains the great significance of aesthetics in human life. This striving to go beyond what is at hand also breeds the creation and appropriation of new forms of aesthetic expressions. The countless opportunities for new forms of expressions that the modern computer enables are just one example of that.

Although art historians, musicologists and others have provided important knowledge on the aesthetic endeavours of humans, we still lack an overarching theory explaining it, as well as lacking the rigorous tools to operationalise such a theory.

5. What are the causes behind the indisputable secular (or even millennial) trend towards declining lethal interpersonal violence in many corners of the world? This seems to also apply to social conflicts whereas it remains to be demonstrated in the case of interstate violence.

It goes without saying that both animals and humans try to avoid getting exposed to lethal violence, which nonetheless endlessly continues to strike unlucky members of all species. But whereas violence among animals shows no tendency to either rise or fall, homicide has gone through an undeniable long-term decline over several centuries, let alone through certain swings up and down. It is beyond reasonable doubt that this decline applies both to interpersonal violence and to violence in social conflicts (although we right now seem to witness a growth of violence in such conflicts) while it is so far less certain that interstate violence has experienced the same downward trend. A number of suggestions regarding how to explain these well-established facts have been offered, but so far none that we find satisfying. In order to get closer to convincing and reliable explanations, a lot of comparative historical research is needed.

6. What role do non-conformist individuals and minority groups play for cultural, cognitive and normative change?

In all times and all societies there exist certain minorities, as well as autonomous individuals, who dissociate themselves from the dominating beliefs and thoughts of their time in different ways – these ideas are represented either by the majority surrounding them or by power-exercising minorities. At the same time, it is quite common that majorities of any time praise the minorities of the past that have eventually come to represent the currently dominant views. This may indicate a wish among the conformist majorities to be more non-conformist than they are or to have the courage to be, but it also indicates that those minorities probably play a pivotal role in the cultural dynamics that characterise human society. Many circumstances suggest that minority groups that are highly dedicated to their cause have a larger impact on the development than their insignificant number might indicate. But the cognitive and normative significance of such groups for cultural change is still unsettled. There is a long-standing tradition of research on conformity and non-conformity, but there is a lack of knowledge about this particular question, like on the risks some people have been willing to take to bring about certain societal changes that would otherwise not have taken place and for which they will only get praised in retrospect.

7. In what way and for what reasons do what humans conceive to be their basic needs and urgent desires change over long periods of time?

The questions of the development of human needs have since long been exiled to the 'department' of the obviously given. This is regrettable, since a lot of aspects pertaining to needs are all but self-evident. In our view, many theories of the hierarchical order of needs, such as Maslow's classical pyramid, insufficiently take into account that needs have a history – that they evolve and change. Without improving theories and methods for the study of the evolution of needs, it will not be possible to really understand the drivers of human history. One essential point of departure would be to make a distinction between needs and wants/desires. *Needs* are what people experience as necessary to fulfil in order to sustain their lives, and they are a subset of their *wants/desires*, which may concern everything from basic necessities, through what are just perceived as necessities onto desires that people know and recognise that they will survive without fulfilling.

The overarching question of needs should be broken down into a number of more precise questions. The first one concerns the change of needs.

8. In what way and for what reasons does everyday sexuality vary as to how it fosters either coherence or discord on the micro level of society as well as on the macro level of society?

Throughout times sexuality has been a highly charged activity. No wonder, then, that literature of all ages and cultures revolve around it as if it were the navel of life. This reflects its highly charged presence in real life (i.e. in human conflicts, whether legally regulated and brought to the court or not). At the same time, it seems quite clear that the family – whatever its composition and structure – has, as a rule, been held together partly by sexual means, never purely aimed at procreation. In addition, it is doubtlessly the case that love and the related attachments between humans have served to keep people together. Our impression is that scientific students of sexuality have been more inclined to address its spectacular expressions and varieties rather than its everyday expressions and varieties, despite the fact that the latter play a significant role in human life as a uniting force no less than as a disruptive force. To really understand sexuality and the paramount role it plays, these unglamorous sides of sex also have to be investigated.

9. In the very long run, the so-called Golden Rule has become more and more generally embraced and yet also intermittently abandoned or denied general applicability. What explains these swings?

The question raised above stems from the following observations:

- Generally, through the ages the idea that all humans are to be considered basically equal has become gradually more widespread and accepted.
- The pace at which the diffusion of this idea has taken place has varied over time.

During certain periods of profound societal transformations, this idea has experienced a lesser degree of acceptance.

It has been claimed by the adherents of the so-called Axial Age theory that over a period of approximately 400–500 years in the past millennium BCE, certain powerful thought systems emerged in different corners of the world.

They were all characterised by a nascent ability among (some intellectually leading) humans to transcend the present state of their societies, as well as themselves and their in-group's self-orientation or self-interest, in favour of considerations of the needs of ever wider segments of humankind. This sea change of thought could be linked to the independent introduction in many places of the Golden Rule. Today, the insistence of even including future generations in such considerations is commonplace – an extension of that rule. However, the Axial Age theory has never been firmly established and is still to be considered a hypothesis. No doubt such thought systems did emerge in the period in question it is, though; far from clear that they were widely and broadly adopted by the peoples of this age. Moreover, whatever their degree of dissemination, through the ages, self-reflective, transcendent thoughts have experienced severe backlashes, which are not taken into account by the Axial Age theory. Thus, here are a set of issues that deserve to be approached, aiming at explaining the ups and downs of the Golden Rule, as well as evaluating the validity of the Axial Age theory.

10. What characterises the evolution of the causal interrelationship between collective solutions, individual solutions and different degrees of individual freedom in society?

In most countries of the contemporary world, one finds an extensive central administration, a tax system and more or less encompassing welfare institutions, as well as a judicial system granting the citizens at least some inalienable human rights. This seems to be the outcome of a secular process leading both to centralisation (e.g. the judicial system) and to decentralisation (e.g. individual human rights). Through taxes (and by other means) the citizens contribute to a set of collective solutions of a number of social and economic problems that were met with other means (if at all) earlier in history. Through certain laws, they are granted individual autonomy in areas of human interaction where in the past they were exposed to disciplining control by the authorities. This is how the trajectory of societal change appears in many societies. Looked upon this way, societal change seems to be a continuous balancing act between collective and individual solutions to problems in human interaction. Perhaps, it is even the case that individualism and collectivism can only evolve in concert. Or are they rather in constant opposition to each other? This needs to be systematically addressed, like the different outcomes of various ways of reconciling individualism and collectivism.

11. Why do some societal processes and phenomena develop in a circular or repetitive way whereas other processes evolve along a cumulative trajectory?

Human interaction is uniquely dynamic in relation to what goes in other parts of the animal world, leading to incessant large- as well as small-scale transitions of our societies. Through the ages the magnitude of human interaction has grown immensely, today being profoundly global in many respects. This also means that an evergrowing number of increasingly differentiated activities are increasingly coordinated, which is synonymous to rising complexity.

These long-term changes are culturally driven in the sense that they do not have any modifications of the human genome as their precondition, despite the likelihood that such modifications evolve nevertheless, although at a slow pace. Many cultural processes are cumulative, which means that they are based on certain experiences which are deployed for further change and so on, in the long run, giving rise to a certain direction of historical changes. This applies to the steadily widening networks of human interaction which today take place on a global scale and to the long-term establishment of a sedentary way of life and so on. On the contrary, other cultural changes are characterised by repetition, sometimes (but far from always) as self-generating cyclical processes. Such recurrences can be seen in most domains of human action as much in the trajectory of ideas (for instance, intelligent design in the twenty-first century resembling the so-called physicotheology of the eighteenth century) or ideologies (e.g. rising and declining xenophobia) and in social or economic development.

Despite the obvious difference between these two types of cultural change, they testify to their common independence of the slowly ongoing genetic modification of humankind. Here, the question at stake is why the cultural dynamics of the human society are enacted either as cumulative change or as recurrent ups and downs. Is it just a matter of innovations that turns out to be functional (propelling cumulative change), as distinct from cultural swings that are functionally neutral (i.e. neither functional nor disfunctional), propelling circular change? Or is it rather the case that recurring phenomena revolve around what really are constant features of human life and human needs? Today we do not have answers to these questions. But they are fruitful questions, inviting us to make systematic inquiries into what characterises and conditions cumulative and circular processes, respectively.

12. Does a gradual extension of our lifespan (and the rise of welfare) imply a growing or declining ability to postpone the satisfaction of our needs and desires?

The ability to postpone rewards or the fulfilment of needs and desires is essential for all human development – both at the individual and at the societal level. We would not survive without it. Thus, it is central to understand what conditions such an ability.

It is reasonable to assume that people need some belief in a good future in order to be willing to trade instant joys for future rewards. Furthermore, one might be inclined to predict that when people's expectation of a long lifespan is low, their incentive to suspend short-term fulfilment might be low too. However, one might well predict the opposite: increased welfare and a longer life will expand the opportunities for instant gratification, which, if so, might reduce the need for postponement and self-control. The question then is: under what cultural (i.e. political, socio-economic/material) conditions are people most ready and capable to discipline their inner cravings for instant satisfaction?

Questions such as these could be addressed in internationally comparative and at the same time historical studies. But they also relate to synchronic-oriented psychological research, such as Walter Mischel's well-known so-called marshmallow experiments testing self-control among children of today. One of the most intriguing results from these experiments is that children who were able to resist their yearning for instant satisfaction tended to have a better career later in life than those who could not resist the temptation to have one marshmallow at once instead of two a little later. Some have traced back the different behavioural responses among the children exposed to this test and see them as reflecting differences in their personalities. But they could just as well be due to differences in trust – those children who thought they had good reasons to believe the promise to get two marshmallows later instead of one now were ready to wait, whereas those who did not trust the adults to the same degree, perhaps because of bad experiences, were less inclined to wait. They might have had a completely rational risk aversion.

In our context Mischel's experiments are relevant, not primarily for what they can tell us about the conditions for individual success but for what they contribute to our understanding of the significance of trust and other circumstances in regard to human's ability and willingness to postpone the satisfaction of their needs and desires.

13. How can the intermittent slowing down of technical change be accounted for in contrast to a general trend towards an ever ongoing speeding up of technical development?

One basic issue that pops up in many of the research questions suggested above is the tension between the stability, or even inertia, and changeability of the human condition. In the domain of technical development this tension comes forth more clearly than elsewhere.

It is indisputable that humankind has generated an enormous technical development throughout its history, which, furthermore, has accelerated through time. Due to well-founded theories, one could claim that this is primarily explained by the continuously growing population of the world accompanied by an incessantly ongoing specialisation of its activities. However, the secular trajectory of accelerating technical change has been disrupted by periods of stagnation many times. No doubt some attempts have been made to come to grips also with periods of retarded technical development. For example, it has been claimed that stagnation, rather than rapid change, is the normal condition due to the infrequent and irregular occurrence of human originality and creativity. However, such an explanation ignores the ever ongoing and everyday small-scale improvements, which often lead to large-scale change in the long run. Another common explanation for the intermittent slowdown is that certain pre-modern societies lacked the institutional preconditions for rapid technical change, which act as incentives for people to pay the cost of invention that, in its turn, is transformed into innovation. But also this so-called property-rights theory may be questioned, again by considering the fact that technical development, be it sometimes almost infinitesimal, has taken place under the most varying circumstances. However, this does not falsify the observation that the history of technical growth has been interleaved with periods of slow or failed growth, evoking the need for theoretical explanations no less than solid empirical studies.

These issues are highly relevant for the citizens of today's societies, comprising large sectors characterised by comparatively slow change. It is a great challenge to identify and analyse the impediments to technical change.

14. What explains the gradual emergence and widespread diffusion of egalitarianism over the centuries?

It is likely that, in the absence of political initiatives, the spontaneous development of society in the long run would lead to a continuously widening gap in the distribution of material resources, in its turn conditioning unequal distribution of other resources (such as health, lifespan and so on) between social classes, as well as between women and men, young and old, people in sparsely populated areas and people in the urban centres. It can be assumed that a small gap in a distant past already gave those with a little more than others an advantage in terms of risk tolerance, leading to a gradually widening gap. However, history has almost never witnessed such an unimpeded spontaneous movement towards evergrowing inequality. To the contrary, in the very long run the general attitude has changed from an unproblematised and often explicit affirmation of inequality towards an increased stress on equality, despite the obvious fact that this has not happened in all places and times over the course of history.

The general evolution of law-making testifies to this sea change. At an early stage legal constraints were put not only on the ordinary subjects but also on the power holders themselves, even within the general framework of inequality before the law. Much later such restrictions were transformed into a universal principle of equality before the law, subsequently followed up by the successive spread of the idea that material sources should be redistributed from the haves to the have-nots.

Of course, the matter of equality remains a contested issue. Nonetheless, the idea of equality seems to prevail in public and private encounters. What is the explanation – a general improvement of welfare? This is far from self-evident. Rising welfare and affluence could trigger the opposite attitude just as well. Thus, here is a field for substantial knowledge advancement.

15. What characterises and explains the linkage between forces that promote social coherence and forces that promote social conflict? Are they mutually exclusive or do they interdependently reinforce each other?

What makes a society hold together, and why does it not fall apart due to all the conflicting interests haunting it? Or is it rather the case that conflicts between social classes and other groups promote the development and unity of society? These are questions that have been asked many times in different ways. However, they are not often addressed together. How are they linked? Perhaps it is precisely through the struggle between people with vested interests that societal coherence evolves and is sustained. Provided that this is the case, it may be linked to growing societal complexity (i.e. the coordination of a gradually growing differentiation of social functions) and probably also to the long-term decline in physical violence. These questions are of paramount importance for our understanding of the structure of human society.

NOTES

I. Questions and Answers – Background, Motivations and Aims

1. Arne Jarrick, *Översyn av humaniora vid Örebro universitet* (Örebro: Örebro universitet, 2015), 11–12.
2. Magnus Bremmer, 'Den frånvarande humanisten kan vitalisera vårt fantasilösa idéklimat', *Dagens Nyheter* 15 March 2019; Isak Hammar, 'Den klassiska folkbildaren kanske tillhör historien', *Dagens Nyheter* 22 March 2019.
3. Poul Holm et al., *The Humanities World Report* (Basingstoke: Palgrave Macmillan, 2015). The quotations do not appear in the book; they are taken from the records of the interviews.
4. Ibid., 55–57.
5. For one example of such an attitude, see Kapil Raj, 'Networks of Knowledge, or Spaces of Circulation? The Birth of British Cartography in Colonial South Asia in Late Eighteenth Century', *Global Intellectual History* 2, no. 1, 2017, 1.
6. See Peter Burke, *What Is the History of Knowledge?* (Cambridge: Polity Press, 2016), 74–76.
7. See Holm et al., *The Humanities World Report*, ch. 2.
8. For two recently published contrasting views on this, see Anna Victoria Hallberg, *Hotet mot humaniora kommer inifrån* (Stockholm: Timbro 2019); Göran Larsson, *Humaniora – en vetenskap bland andra* (Stockholm: Timbro, 2019).
9. C. P. Snow, *The Two Cultures* (London: Cambridge University Press, 1993).
10. This is the overarching message in *Forskningsöversikt 2019: Humaniora och samhällsvetenskap* (Stockholm: Vetenskapsrådet, 2019).
11. This was, though, highly controversial in the 1920s and 1930s – see Thord Silverbark, *Fysikens filosofi: Diskussioner om Einstein, relativitetsteorin och kvantfysiken i Sverige 1910—1970* (Stockholm: Brutus Östlings förlag, 1999).
12. Best described in Ernst Mayr, *The Growth of Biological Thought: Diversity, Evolution, and Inheritance* (Cambridge, MA: Belknap Press of Harvard University Press, 1982), esp. ch. 12.
13. Nicholas Jardine & Emma Spary, 'Worlds of History', in H. A. Curry et al. (eds), *Worlds of Natural History* (Cambridge: Cambridge University Press 2018), 8–9; Burke *What Is the History of Knowledge?*, 6.
14. Klas Åmark, *Att bo granne med ondskan: Sveriges förhållande till nazismen, Nazityskland och förintelsen* (Stockholm: Bonniers, 2011).
15. Manuel Eisner, 'From Swords to Words: Does Macro-Level Change in Self-Control Predict Long-Term Variation in Levels of Homicide?', *Crime and Justice* 33, no. 1, 2014, 1–72.

16 Nuria Ristin-Kaufman & Marianne Gullberg, 'The Effects of First Exposure to an Unknown Language at Different Ages', *Bulletin, Suisse de Linguistique Appliquée* 99, 2014, 17–29.
17 Lars Nyberg et al., 'Memory Aging and Brain Maintenance', *Trends in Cognitive Science* 16, no. 5, 2012, 292–305.
18 Gustaf Gredebäck & Annika Melinder, 'Infants' Understanding of Everyday Social Interactions: A Dual Process Account', *Cognition* 114, no. 2, 2010.
19 Michael Tomasello, *The Cultural Origins of Human Cognition* (Cambridge, MA: Harvard University Press 1999).
20 Patrik Lindenfors et al., 'The Cultural Evolution of Democracy: Saltational Changes in a Political Regime Landscape', *PLoS One* 6, no. 11, 2011; Fredrik Jansson et al., 'Democratic Revolutions as Institutional Innovation Diffusion: Rapid Adoption and Survival of Democracy', *Technological Forecasting & Social Change* 80, 2013.
21 Arne Jarrick & Janken Myrdal, 'De stora forskningsfrågorna om människans villkor', *Respons* 5, 2015.
22 Arne Jarrick, 'Knowledge Resistance: A Global Challenge – in Research and Education, in the Humanities and Elsewhere', in *Humanities and Higher Education: Synergies between Science, Technology and Humanities (Higher Education in the World 7)* (Barcelona: GUNi network, 2019).

Chapter 1. "What Explains That Some Kinds of Knowledge Are Widely Accepted Whereas Other Kinds of Knowledge Are Rejected?

1 For another perspective, see Peter Burke, *What Is the History of Knowledge?* (Cambridge: Polity Press, 2016), 7–9, 114.
2 Ibid., 113–14.
3 Charles G. Lord et al., 'Biased Assimilation and Attitude Polarization: The Effects of Prior Theories on Subsequently Considered Evidence', *Journal of Personality and Social Psychology* 37, no. 11, 1979, 2098–109.
4 Parts of this section have been published in Arne Jarrick, 'Knowledge Resistance: A Global Challenge – in Research and Education, in the Humanities and Elsewhere', in David Bueno et al. (eds), *Humanities and Higher Education: Synergies between Science, Technology and Humanities, Higher Education in the World 7* (Barcelona: GUNi network, 2019), 77–84. Also see Burke, *What Is the History of Knowledge?*, 24–25.
5 Robert Lane, 'The Decline of Politics and Ideology in a Knowledgeable Society', *American Sociological Review* 31, no. 5, 1965, 649–62; Alain Touraine, *The Postindustrial Society: Tomorrow's Social History: Classes, Conflicts and Culture in the Programmed Society* (New York: Random House, 1971); Peter F. Drucker, *The Age of Discontinuity: Guidelines to Our Changing Society* (London: Harper & Row, 1969).
6 Daniel Bell, *The Coming of the Post-Industrial Society: A Venture in Social Forecasting* (New York: Basic Books, 1973).
7 See, for example, Gernot Böhme & Nico Stehr (eds), *Sociology of the Sciences Yearbook 1986: The Knowledge Society* (Dordrecht: D. Reidel, 1986).
8 Auguste Comte, *A General View of Positivism* (London: Trübner, 1865); Alfred Whitehead, *Science and the Modern World* (New York: Macmillan, 1925).
9 Some observations indicate that positive attitudes to knowledge and science are deteriorating. See, for example, Soroush Vosoughi et al., 'The Spread of True and

False News Online', *Science* 359, no. 6380, 2018, 1146–51; Hunt Allcott & Matthew Gentzkow, 'Social Media and Fake News in the 2016 Election', *Journal of Economic Perspectives* 31, no. 2, 2017, 211–36. They are, though, still not conclusive.

10 The following nine paragraphs are based on Arne Jarrick, *Det finns inga häxor: en bok om kunskap* (Stockholm: Weyler, 2017).

11 James R. Flynn, 'Massive IQ Gains in 14 Nations: What IQ Tests Really Measure', *Psychological Bulletin* 101 no. 2, 1987, 171–91.

12 For example, OECD, *Literacy in the Information Age* (Paris: OECD, 2000); OECD, *Skills Outlook 2013* (Paris: OECD, 2013); Burke *What Is the History of Knowledge?*, 94–96.

13 Augusto López-Claros & Bahiyyih Nakhjavani, *Equality for Women = Prosperity for All* (New York: St. Martin's Press, 2018), ch. 6.

14 About autocratisation in general: V-Dem, 'The Third Wave of Autocratization is Here', 11 March 2019, https://www.v-dem.net/en/news/third-wave-autocratization-here/; about Greece: IAOS, '80 Former Chief Statisticians Condemn Prosecution of Andreas Georgiou', 18 June 2018, https://www.iaos-isi.org/index.php/latestnews/221-80-former-chief-statisticians-condemn-prosecution-of-andreas-georgiou; about Brazil: Doris Elin Urrutia, 'Brazil's Space Agency Head Was Forced Out for Defending Climate Science', 18 June 2018, https://www.space.com/brazil-space-agency-leader-dismissed.html; about Hungary: Arne Jarrick & Peter Pagin, 'Democracy and Academic Freedom – the Hungarian Example', 10 December 2019, https://kva.se/en/nyheter/demokrati-och-akademisk-frihet-exemplet-ungern, https://kva.se/en/nyheter/demokrati-och-akademisk-frihet-exemplet-ungern and the like.

15 Lars Calmfors, 'Migration, sysselsättning och löner. Hur ska vi ta itu med arbetsmarknadens Tudelning', in Johan Lundberg (ed.), *Immigration, ekonomisk integration och entreprenörskap* (Stockholm: Swedish Economic Forum, 2016).

16 Michael Marmot, *Status Syndrome: How Your Social Standing Directly Affects Your Health and Life Expectancy* (London: Bloomsbury, 2004); Nordforsk, *Adding Value to Nordic Research Cooperation: Recommendations for a New Initiative on Distribution of Health* (Oslo: Nordforsk, 2012); Robert Eriksson, 'Why Do Graduates Live Longer?', in J. O. Jonsson & C. Mills (eds), *Cradle to Grave: Life-Course in Modern Sweden* (Durham: Sociology Press, 2001); Silvia Stringhini et al., 'Socioeconomic Status and the 25 × 25 Risk Factors as Determinants of Premature Mortality', *Lancet*, no. 389, 2017, 1229–37, http://www.thelancet.com/pdfs/journals/lancet/PIIS0140-6736(16)32380-7.pdf; Malavika Subramanyam et al., 'Relative Deprivation in Income and Self-Rated Health in the United States', *Social Science & Medicine* 69, no. 3, 2009, 327–34.

17 Rune Åberg, 'Unemployment Persistency. Over-Education and the Employment of the Less Educated', *European Sociological Review* 19, no. 2, 2003, 199–216.

18 K. A. Ericsson, 'An Introduction to *The Cambridge Handbook of Expertise and Expert Performance: Its Development, Organization, and Content*', in K. A. Ericsson et al. (eds), *The Cambridge Handbook of Expertise and Expert Performance* (Cambridge: Cambridge University Press, 2006), 6, 3–20.

19 Georges Ifrah, *The Universal History of Numbers: From Prehistory to the Invention of the Computer, Part 1* (London: Harvill Press, 1998), 278–80, 339–40.

20 For an overview, see Burke, *What Is the History of Knowledge?*, esp. 77–80; also see Johan Östling et al., 'The History of Knowledge and the Circulation of Knowledge: An Introduction', in Johan Östling et al. (eds), *Circulation of Knowledge: Explorations in the History of Knowledge* (Lund: Nordic Academic Press, 2018), 9–33; compare with Andreas W. Daum, 'Varieties of Popular Science and the Transformation of Public Knowledge: Some Historical Reflections', *Isis* 100, no. 2, 2009, 331.

21 See, for example, James Secord, 'Knowledge in Transit', *Isis* 95, no. 4, 2004; Östling et al. (eds), *Circulation of Knowledge*; Anders Ekström (ed.), *Den mediala vetenskapen* (Nora: Nya Doxa, 2004), see most chapters, esp. Ekström's introduction, 9–31, and Kjell Jonsson & Jenny Larsson, 'Mellan vetenskap och vers: Ärftlighetsforskare som publika vetenskapsmän 1940–1960', in Anders Ekström (ed.), Den mediala vetenskapen (Nora: Nya Doxa, 2004), 220.
22 Burke *What Is the History of Knowledge?*, 4, 6.
23 For example, Daum 'Varieties of popular science …', 324–26; Catherine Jami, 'The Jesuits' Negotiation of Science between France and China (1685–1722): Knowledge and Modes of Imperial Expansion', in Lázló Kontler et al. (eds), *Negotiating Knowledge in Early-Modern Empires: A Decentered View* (Palgrave Macmillan: New York, 2014), 53–77; Kapil Raj, 'Networks of Knowledge, or Spaces of Circulation? The Birth of British Cartography in Colonial South Asia in the Late Eighteenth Century', *Global Intellectual History* 2, no. 1, 2017, 49–66; Burke, *What Is the History of Knowledge?*, 5, 83–84.
24 Roger Chartier, *Cultural History: Between Practices and Representations* (Cambridge: Polity in association with Blackwell, 1988); Carlo Ginzburg, *The Cheese and the Worms: The Cosmos of a Sixteenth-Century Miller* (London: Routledge, 1980).
25 See, for example, Lissa Roberts, 'Situating Science in Global History', *Itinerario* 33, no. 1, 2009, 15, 21; Burke, *What Is the History of Knowledge?*, 83–85, 87.
26 Simon Schaffer, 'Newton on the Beach: The Information Order of *Principia Mathematica*', *History of Science* 47, no. 3, 2009, 243–76.
27 For two examples of that, see Laura Hollsten, 'Public, Private, and Experience-Based Knowledge: Cholesterol Knowledge in Circulation in Finnish Society, 1970–2010', in Johan Östling et al. (eds), *Circulation of Knowledge: Explorations in the History of Knowledge* (Lund: Nordic Academic Press, 2018); and Thord Silverbark, *Fysikens filosofi: Diskussioner om Einstein, relativitetsteorin och kvantfysiken i Sverige 1910—1970* (Stockholm/Stehag: Brutus Östlings förlag, 1999).
28 David Wootton, *The Invention of Science: A New History of the Scientific Revolution* (New York: Harper Collins, 2015), 78; Dava Sobel, *Galileo's Daughter: A Drama of Science, Faith, and Love* (London: Fourth Estate, 1999), 116. Also see Noah J. Efron, 'Myth 9. That Christianity Gave Birth to Modern Science', in Ronald L. Numbers (ed.), *Galileo Goes to Jail and Other Myths about Science and Religion* (Cambridge, MA: Harvard University Press, 2009), 81–83.
29 Lindsey Fitzharris, *The Butchering Art: Joseph Lister's Quest to Transform the Grisly World of Victorian Medicine* (New York: Scientific American/Farrar Straus and Giroux, 2017), 196–98.
30 Charles Darwin, *Självbiografi* (Stockholm: Natur och kultur, 2009), 133.
31 See Wootton, *The Invention of Science*, for a recent very impressive study.
32 Lord et al., 'Biased Assimilation and Attitude Polarization'.
33 K. Edwards & E. E. Smith, 'A Disconfirmation Bias in the Evaluation of Arguments', *Journal of Personality and Social Psychology*, 71, no. 1, 1996; Charles S. Taber & Milton Lodge, 'Motivated Skepticism in the Evaluation of Political Beliefs', *American Journal of Political Science* 50, no. 3, 2006; Martin Gilens, 'Political Ignorance and Collective Policy Preferences', *American Political Science Review* 95, no. 2, 2001.
34 Dan Kahan et al., 'Motivated Numeracy and Enlightened Self-Government', *Behavioural Public Policy* 1, no. 1, 2017, 54–86; Brendan Nyhan & Jason Reifler, 'When Corrections Fail: The Persistence of Political Misperceptions', *Political Behavior*

32, no. 2, 2010, 303–30; Brendan Nyhan et al., 'Effective Messages in Vaccine Promotion: A Randomized Trial', *Pediatrics* 133, no. 4, 2015, e835–e842; Gordon Pennycook & David G Rand, 'Who Falls for Fake News? The Roles of Bullshit Receptivity, Overclaiming, Familiarity, and Analytic Thinking', *Journal of Personality* 88, no. 2, 2019, 185–200.
35   Brendan Nyhan et al., 'Taking Corrections Literally but Not Seriously? The Effects of Information on Factual Beliefs and Candidate Favorability', https://papers.ssrn.com/sol3/papers.cfm?abstract_id=2995128; Briony Swire et al., 'Processing Political Misinformation: Comprehending the Trump Phenomenon', *Royal Society Open Science* 1 March, 2017, https://royalsocietypublishing.org/doi/full/10.1098/rsos.160802.
36   See, for example, Robert C. Luskin et al., 'Considered Opinions: Deliberative Polling in Britain', *British Journal of Political Science* 32, no. 3, 2002, 455–87; also see John Cook & Stephan Lewandowsky. 'The Debunking Handbook', 27 November 2011, https://skepticalscience.com/Debunking-Handbook-now-freely-available-download.html; Douglas Guilbeault et al., 'Social Learning and Partisan Bias in the Interpretation of Climate Trends', *PNAS* 115, 2018, 9714–19.
37   Foremost among researchers doing this is Daniel Kahneman, *Thinking, Fast and Slow* (New York: Farrar, Straus and Giroux, 2013); Peter Turchin et al., 'Quantitative Historical Analysis Uncovers a Single Dimension of Complexity That Structures Global Variation in Human Social Organization', *PNAS*, 9 January 2018, https://www.pnas.org/content/115/2/E144.short.
38   Riksbankens Jubileumsfond. Åsa Wikforss. 'Knowledge Resistance: Causes, Consequences, Cures', https://www.rj.se/en/anslag/2018/knowledge-resistance-causes-consequences-and-cures/. It is funded by Riksbankens Jubileumsfond, based in Sweden.

Chapter 2. Why Do Some Societal Processes and Phenomena Develop in a Circular or Repetitive Way Whereas Other Processes Evolve along a Cumulative Trajectory?

1   Stefano Ghirlanda et al., 'Memory for Stimulus Sequences: A Divide between Humans and Other Animals?', *Royal Society Open Science*, 21 June 2017, https://royalsocietypublishing.org/doi/full/10.1098/rsos.161011.
2   Compare with Joseph Tainter, *The Collapse of Complex Societies* (Cambridge: Cambridge University Press, 1988); Edgar Morin, *On Complexity* (Cresskill, NJ: Hampton Press, 2008); Herbert O. Simon, 'The Architecture of Complexity', In *Systems Thinking (vol. 1)*, edited by Gerald Midgley. London: Sage, 2003, 306–408.
3   On this, see Arne Jarrick & Maria Wallenberg Bondesson, *The Dynamics of Law-Making: A World History* (Stockholm: KVHAA 2018), 34–39.
4   Matt Ridley, *Nature via Nurture: Genes, Experience & What Makes Us Human* (New York: HarperCollins 2003), 35–37.
5   On this, see Arne Jarrick et al. (eds), *Methods in World History: A Critical Approach* (Lund: Nordic Academic Press, 2016), 'Introduction'.
6   Arne Jarrick, *The Need to Be Needed: An Essay on Humankind, Culture, and World History* (Saarbrücken: Lambert Academic, 2010), 78–90.
7   Jarrick & Wallenberg Bondesson, *Dynamics of Law-Making*, 347.

8   Compare with Roger Scruton, *Beauty: A Very Short Introduction* (Oxford: Oxford University Press, 2009), 142–43; Ronald Sanders, *The Days Grow Short: The Life and Music of Kurt Weill* (New York: Holt, Rinehart and Winston, 1980), 27, 47; Wikipedia, 'Igor Stravinskij', https://sv.wikipedia.org/wiki/Igor_Stravinskij.
9   About the change of the occupational structure, see Arne Jarrick, *The Need to Be Needed*, 145–48; Robert Fogel, *The Fourth Great Awakening & the Future of Egalitarianism* (Chicago: University of Chicago Press, 2000), ch. 5; Tommy Isidorsson, *Striden om tiden: Arbetstidens utveckling i Sverige under 100 år i ett internationellt perspektiv* (Göteborg: Department of History, 2001), ch. 4; Alf Johansson, *Den effektiva arbetstiden: Verkstäderna och arbetsintensitetens problem 1900–1920* (Uppsala: Uppsala University, 1977); Lennart Schön, *En modern svensk ekonomisk historia* (Stockholm: SNS, 2007), 301, 376; John R. McNeill & William McNeill, *The Human Web: A Bird's-Eye View of World History* (New York: W. W. Norton, 2003), 311.
10 Sten Lindroth, *Svensk lärdomshistoria: Gustavianska tiden* (Stockholm: Norstedts, 1989).
11 See, for instance, Thomas Piketty, *Capital in the Twenty-First Century* (Cambridge, MA: Belknap Press of Harvard University Press, 2014); Per Molander, *The Anatomy of Inequality* (Brooklyn: Melville House Books, 2017); Rolf Aaberge et al. (eds), *Increasing Income Inequality in the Nordics* (*Nordic Economic Policy Review*, 2018); Branko Milanovic, *Global Inequality: A New Approach for the Age of Globalization* (Cambridge, MA: Harvard University Press, 2016); Peter Turchin & Sergey A Nefedov, *Secular Cycles* (Princeton, NJ: Princeton University Press, 2009); Simon Kuznets, 'Economic Growth and Income Inequality', *American Economic Review* 45, no. 1, 1955.
12 Jarrick & Wallenberg Bondesson, *The Dynamics of Law-Making*.
13 Peter Turchin, *War and Peace and War: The Rise and Fall of Empires* (London: Penguin books, 2006); also see Albert O. Hirschman, *Shifting Involvements: Private Interest and Public Action* (Princeton, NJ: Princeton University Press 2002), for an ingenious idea about self-generating mental shifts.
14 John M. Keynes, *The General Theory of Employment, Interest and Money* (London: Macmillan, 1973), ch. 22; Paul Samuelson, 'Synthesis of the Principle of Acceleration and the Multiplier', *Journal of Political Economy*, 47, no. 6, 1939, 786–97.
15 Schön, *En modern svensk ekonomisk historia*.
16 Johan Galtung & Sohail Inayatullah, *Macrohistory and Macrohistorians: Perspectives on Individual, Social, and Civilizational Change* (Westport, CT: Praeger, 1997).
17 Fernand Braudel, *The Mediterranean and the World in the Age of Philip II* (London: Collins, 1972–73); Fernand Braudel, 'History and the Social Sciences: The *Longue Durée*', in Fernand Braudel (ed.), *On History* (Chicago: University of Chicago Press, 1980).
18 Compare Pitirim Sorokin's distinction between inward- and outward-looking societies and times. Pitirim Sorokin, *Social and Cultural Dynamics: A Study of Change in Major Systems of Art, Truth, Ethics, Law and Social Relationships* (Boston, MA: Porter Sargent, 1957).
19 See, for instance, Galtung & Inayatullah, *Macrohistory and Macrohistorians*.
20 Wootton, *The Invention of Science: A New History of the Scientific Revolution* (New York: HarperCollins, 2015), 565. This book abounds with such examples.
21 Michael H. Westbrook, *The Electric Car: Development and Future of Battery, Hybrid and Fuel-Cell Cars* (London: Institution of Electric Engineers, 2001), 15.
22 David V. Herlihy, *Bicycle: The History* (New Haven, CT: Yale University Press, 2004).
23 Pierre Lemonnier, 'introduction', in Pierre Lemonnier (ed.), *Technological Choices: Transformation in Material Cultures since the Neolithic* (London: Routledge, 1993), 24–26.

24 Georges Guille-Escuret, 'Technical Innovation and Cultural Resistance', in Pierre Lemonnier (ed.), *Technological Choices: Transformation in Material Cultures since the Neolithic*; Lemonnier, 'Introduction', 17.
25 Pierre Pétrequin, 'North Wind, South Wind: Neolithical Technical Choices in the Jura Mountains, 3700–2400 BC'. ...', in Pierre Lemonnier (ed.), *Technological Choices: Transformation in Material Cultures since the Neolithic*, 53.
26 Sander van der Leeuw, 'Giving the Potter a Choice', in Pierre Lemonnier (ed.), *Technological Choices: Transformation in Material Cultures since the Neolithic*, 238–83.
27 Compare with Lemonnier, 'Introduction', 23.
28 Jarrick & Wallenberg Bondesson, *Dynamics of Law-Making*.

Chapter 3. Why Do Social Norms Change, Despite the Fact That Their Mission Is to Be Sustained? What Role Do Non-Conformist Individuals and Minority Groups Play in Cultural, Cognitive and Normative Change?

1 Pontus Strimling et al., 'The Connection between Moral Positions and Moral Arguments Drives Opinion Change', *Nature Human Behaviour* 3, 15 July, 2019, 922–30.
2 Manuel Eisner, 'From Swords to Words: Does Macro-Level Change in Self-Control Predict Long-Term Variation in Levels of Homicide?', *Crime and Justice* 43, no. 1, 2014, 1–72.
3 This is what Steven Pinker claims (on shaky grounds) in *The Better Angels of Our Nature: The Decline of Violence in History and Its Causes* (London: Penguin, 2011), ch. 5.
4 Compare with Arne Jarrick, *Back to Modern Reason: Johan Hjerpe and Other Petit Bourgeois in Stockholm in the Age of Enlightenment* (Liverpool: Liverpool University Press, 1999), ch. 2.
5 Compare with Arne Jarrick & Johan Söderberg, 'Inledning', in Arne Jarrick & Johan Söderberg (eds), *Människovärdet och makten* (Stockholm: Stockholmia förlag, 1994).
6 Strimling et al., 'The Connection between Moral Positions and Moral Arguments Drives Opinion Change' (2019), 922–30.
7 See Damon Centola et al., 'Experimental Evidence for Tipping Points in Social Convention', *Science* 360, no. 6393, 2018, 1116–19.
8 Laurence Steinberg & Kathryn C. Monahan, 'Age Differences in Resistance to Peer Influence', *Developmental Psychology* 43, no. 6, 2007, 12: 'After all, individual conformity in any situation is the joint product of the pressure to conform and the capacity of the individual to resist the coercion.'
9 Some of the classic studies on the tendency to conform are S. E. Asch, *Social Psychology* (New York, Prentice-Hal, 1952); Stanley Milgram, *Obedience to Authority, An Experimental View* (London: Pinter & Martin 2010); Graig Haney et al., 'Interpersonal Dynamics in a Simulated Prison', *International Journal of Criminology and Penology* 1, no. 1, 1973, 69–97. For a discussion of Milgram's status, see Thomas Blass (ed.), *Obedience to Authority: Current Perspectives on the Milgram Paradigm* (Mahwah, NJ: Lawrence Erlbaum Associates, 2000), although this was published before Haslam and Reichel's critical experiments. For the statement on the dominance and long tradition of conformity studies, see Serge Moscovici, 'Social Influence, Conformity Bias, and the Study of Active Minorities', in Leonard Berkowitz (ed.), *Advances in Experimental Psychology, Vol. 6* (New York: Academic Press, 1972), 150–51; Vladas Girskevicius et al., 'Going Along versus Going Alone: When Fundamental Motives Facilitate Strategic (Non) Conformity', *Journal of Personality and Social Psychology* 91, no. 2, 2006, 281–82.

10 Rod Bond & Peter J. Smith, 'Culture and Conformity: A Meta-Analysis of Studies Using Asch's (1952b, 1956) Line Judgment Task', *Psychological Bulletin*, 119, no. 1, 1996, 111–37.
11 Muzafer Sherif, *The Psychology of Social Norms* (New York: Harper & Bros, 1936), especially chs V–VI, where he established in the latter his renowned autokinetic effect. Sherif in turn referred critically to F. H. Allport's studies on the subject, 69, 81–82.
12 My interpretation is influenced by conversations with Torun Lindholm, professor of psychology at Stockholm University. See Asch, *Social Psychology*, 407, 484–85, for an outline of Sherif's experiments. Sherif deliberately arranged his experiments to be 'objectively unstable', Sherif, *The Psychology of Social Norms*, 91, 107.
13 Cited from Bond & Smith, 'Culture and Conformity', 111.
14 The distinction was established by M. Deutsch and H. B. Gerard. On this subject, see Hyung (David) Oh, 'Do Collectivists Conform More than Individualists? Cross-Cultural Differences in Compliance and Internalization', *Social Behavior and Personality* 41, no. 6, 2013, 982.
15 The distinction was established by L. Festinger. See Bond & Smith 'Culture and Conformity', 128.
16 Examples of the new orientation: see Dominic J. Packer, 'On Being Both with Us and Against Us: A Normative Conflict Model of Dissent in Social Groups', *Personality and Social Psychology Review*, 12, no. 1, 2008, 50–71; Geoff MacDonald et al., 'Do People Use Reverse Psychology? An Exploration of Strategic Self-Anticonformity', *Social Influence*, 6, no.1, 2011, 1–14; Matthew Hornsey et al., 'On Being Loud and Proud: Non-Conformity and Counter-Conformity to Group Norms', *British Journal of Social Psychology*, 42, no. 3, 2003, 319–35; Moscovici, 'Social Influence, Conformity Bias, and the Study of Active Minorities'.
17 See, for example, Alexander S. Haslam & Stephen D. Reicher, 'The Psychology of Tyranny', *Scientific American Mind* 16, no. 3, 2005, 44–51; S. Alexander Haslam & Stephen Reicher, 'Beyond the Banality of Evil: Three Dynamics of an Interactionist Social Psychology of Tyranny', *Personality and Social Psychology Bulletin* 33, no. 5, 2007, 615–22; Stephen Reicher & S. Alexander Haslam, 'On the Agency of Individuals and Groups: Lessons from the BBC Prison Study', in Tom Postmes & Jolanda Jetten (eds), *Individuality and the Group: Advances in Social Identity* (London: Sage, 2006), 237–57, for an attempt at empirical criticism of Zimbardo's prisoner/guard experiment. For Zimbardo's critical response, see Philip G. Zimbardo, 'On Rethinking the Psychology of Tyranny: The BBC Prison Study', *British Journal of Social Psychology* 45, no. 1, 2006, 1–40. For an overview of relevant research in social neuroscience, see Keise Izuma, 'The Neural Basis of Social Influence and Attitude Change', *Current Opinion in Neurobiology* 23, no. 3, 2013, 456–62.
18 Thomas Blass, 'The Milgram Paradigm after 35 Years: Some Things We Now Know about Obedience to Authority', in Blass (ed.), *Obedience to Authority*, 50–52.
19 Bond & Smith, 'Culture and Conformity', 121; Stephen J. Cutler & Robert L. Kaufman, 'Cohort Changes in Political Attitudes: Tolerance of Ideological Nonconformity', *Public Opinion Quarterly* 39, no. 1, 1975, 69–81.
20 Bond & Smith, 'Culture and Conformity', 115, 117–18, 125; Haney et al., 'Interpersonal Dynamics in a Simulated Prison', 73.
21 Examples: Sherif, *The Psychology of Social Norms*, 95; Girskevicius et al., 'Going Along versus Going Alone'; Milgram, *Obedience to Authority*; Dianne M. Tice, 'Self-Concept Change and Self Presentation: The Looking Glass Self Is also a Magnifying Glass', *Journal of Personality and Social Psychology* 63, no. 3, 1992, 435–51; for an interesting exception, see Steinberg & Monahan, 'Age Differences in Resistance to Peer Influence'.

22 For example, Bond & Smith, 'Culture and Conformity', 122; Bert H. Hodges & Anne L. Geyer, 'A Nonconformist Account of the Asch Experiments: Values, Pragmatics, and Moral Dilemmas', *Personality and Social Psychology Review* 10, no. 1, 2006, 7, 11; Steinberg & Monahan, 'Age Differences in Resistance to Peer Influence', 4, 6, 8–10.
23 See, for example, Oh, 'Do Collectivists Conform More than Individualists? Cross-Cultural Differences in Compliance and Internalization'; Marco Cinnirella & Ben Green, 'Does "Cyber-Conformity" Vary Cross-Culturally? Exploring the Effect of Culture and Communication Medium on Social Conformity', *Computers in Human Behavior* 23, no. 4, 2007, 2011–25; Steinberg & Monahan, 'Age Differences in Resistance to Peer Influence', 11. Much of this is based on the studies of Harry C. Triandis, including his 'The Self and Social Behavior in Differing Cultural Contexts', *Psychological Review* 96, no. 3, 1989, especially 509–10, 513, and his 'Individualism–Collectivism and Personality', *Journal of Personality* 69, no. 6, 2001, 907–24. References for how societies diverge along the dimension of individualism–collectivism are often given to Geert Hofstede. See his overview article, 'The Cultural Relativity of organizational Practices and Theories', *Journal of International Business Studies* 14, no. 2, 1983, 75–89.
24 Michael H. Logan & Hector N. Qirko, 'An Evolutionary Perspective on Maladaptive Traits and Cultural Conformity', *American Journal of Human Biology* 8, no. 5, 1996, 625–26.
25 Hodges & Geyer, 'A Nonconformist Account of the Asch Experiments'.
26 Hodges & Geyer, 'A Nonconformist Account of the Asch Experiments', 3–4. Asch also realised this; see Asch, *Social Psychology*, 14. On the number of studies, see Bond & Smith, 'Culture and Conformity', 118.
27 See, for example, Reicher & Haslam, 'On the Agency of Individuals and Groups'. Compare with Sherif, *The Psychology of Social Norms*, 70–76, for a balanced view.
28 Christopher Browning, *Ordinary Men: Reserve Police Battalion 101 and the Final Solution in Poland* (New York: HarperCollins, 1992).
29 Haney et al., 'Interpersonal Dynamics in a Simulated Prison', 89–91.
30 The idea of the banality of evil was coined by Hanna Arendt when she witnessed the trial of Adolf Eichmann in Jerusalem. See her book *Eichmann in Jerusalem: A Report on the Banality of Evil* (New York: Viking Press, 1964), especially Chapters 2 and 7.
31 Arendt, *Eichmann in Jerusalem*, 276.
32 Haslam & Reicher, 'Beyond the Banality of Evil', 616. Compare with Haney et al., 'Interpersonal Dynamics in a Simulated Prison', 70–73.
33 See Logan & Qirko, 'An Evolutionary Perspective on Maladaptive Traits…', 625, and the references cited.
34 Cutler & Kaufman, 'Cohort Changes in Political Attitudes', 79.
35 S. P. Oliner & P. M. Oliner, *The Altruistic Personality – Rescuers of Jews in Nazi Europe: What Led Ordinary Men and Women to Risk Their Lives on Behalf of Others?* (New York: Free Press, 1988). The book is however focused on altruism rather than resistance to group pressure. For understandable reasons, rescuers must work covertly, even if they protested openly in some cases, as in the Netherlands, 51, 80. Compare with François Rochat & Andre Modigliani, 'Captain Paul Grueninger: The Chief of Police Who Saved Jewish Refugees by Refusing to Do His Duty', in Blass (ed.), *Obedience to Authority*, 100, where the obstruction was overt.
36 Compare with Haslam & Reicher, 'Beyond the Banality of Evil', 621: 'the critical issue is no longer who goes along with brutality (and why) but rather why certain individuals are exceptional enough to oppose it.'

37 For orientation regarding this, see Arne Jarrick & Maria Wallenberg Bondesson, 'Flexible Comparativeness: Towards Better Methods for the Cultural Historical Study of Laws and Other Aspects of Human Culture', in Anna Maria Forssberg et al. (eds), *Organizing History: Studies in Honour of Jan Glete* (Lund: Nordic Academic Press, 2011), 179–99.
38 Triandis, 'The Self and Social Behavior in Differing Cultural Contexts'; Harry C. Triandis, 'Individualism – Collectivism and Personality', *Journal of Personality* 69, no. 6, 2001; Oh, 'Do Collectivists Conform More than Individualists?'.
39 Psychological research supports the notion that this applies to the influence of majorities in general. Serge Moscovici has shown this, cited in Bond & Smith, 'Culture and Conformity', 113.
40 H. L. A. Hart, *The Concept of Law* (Oxford: Oxford University Press, 1997), 23, 50–71, 92. A fundamental idea in Niklas Luhmann, *Law as a Social System* (Oxford: Oxford University Press, 2004).
41 Rochat & Modigliani, 'Captain Paul Grueninger', 100.
42 Albert O. Hirschman, *Exit, Voice and Loyalty: Responses to Decline in Firms, Organizations, and States* (Cambridge, MA: Harvard University Press, 1970), 70; Moscovici, 'Social Influence, Conformity Bias, and the Study of Active Minorities'. The argument also recurs in Matthew J. Hornsey, 'Ingroup Critics and Their Influence on Groups', in Postmes & Jetten (eds), *Individuality and the Group*, 74–75 and the entire article; Reicher & Haslam, 'On the Agency of Individuals and Groups', 249; Packer, 'On Being Both with Us and Against Us'.
43 Arnold J. Toynbee, *A Study of History: Abridgement of Volumes I-VI by D.C. Somervell* (Oxford: Oxford University Press, 1974), 230–45.
44 Logan & Qirko, 'An Evolutionary Perspective on Maladaptive Traits…', 622, 624, quotation on 624.
45 Robert Boyd & Peter J. Richerson, *Culture and the Evolutionary Process* (Chicago: University of Chicago Press, 1985), 213–23.

Chapter 4. Does a Gradual Extension of Our Lifespan (and the Rise of Welfare) Imply a Growing or Declining Ability to Postpone the Satisfaction of Our Needs and Desires?

1 Roger Scruton, *Beauty: A Very Short Introduction* (Oxford: Oxford University Press, 2009), 46.
2 For example, see Joseph Tainter, *The Collapse of Complex Societies* (Cambridge: Cambridge University Press, 1988); also see Daniel Dorling, *Injustice: Why Social Inequality Persists* (Bristol: Policy Press, 2010), 135–37, 211.
3 Robert Fogel, *The Fourth Great Awakening and the Future of Egalitarianism* (Chicago: University of Chicago Press, 2000), 143. Here strong evidence is presented that this applies to the British people, for example.
4 On this, see, for example, Dorling, *Injustice*, 134, ch. 6 (e.g. 212–15).
5 For some aspects of this, see Michael Marmot, *Status Syndrome: How Your Social Standing Directly Affects Your Health and Life Expectancy* (London: Bloomsbury, 2004).
6 Fogel, *The Fourth Great Awakening*, 191 and the rest of ch. 5.
7 Thomas Piketty, *Capital in the Twenty-First Century* (Cambridge, MA: Belknap Press of Harvard University Press, 2014), 324, 327; Branko Milanovic, *The Haves and Have-nots* (New York: Basic Books, 2016), 100–101.

8   Based on John R. McNeill, *Something New under the Sun: An Environmental History of the Twentieth-Century World* (London: Penguin Books, 2000), ch. 1. Also see Hilde Ibsen, '*Global økologisk historie*', in Arne Jarrick & Alf Johansson (eds), *Jorden runt igen – nya bidrag till en gammal globalhistoria* – 25:e Nordiska historiemötet 2004, Stockholm: Almqvist & Wiksell International, 2004), 77–78.
9   For example, see Emmanuel Le Roy Ladurie, 'A Great Agrarian Cycle', *The Peasants of Languedoc* (Urbana: University of Illinois, 1974), 289–311. It was originally published in 1966.
10 Christopher Lasch, *The Culture of Narcissism: American Life in an Age of Diminishing Expectations* (New York: W.W. Norton, 1978). Compare with Charles Taylor, *The Ethics of Authenticity* (Cambridge, MA: Harvard University Press, 1991), esp. ch. 6.
11 Dorling, *Injustice: Why Social Inequality Persists*, 253.
12 Ibid., 248.
13 For example, see Peter Turchin, *War and Peace and War: The Rise and Fall of Empires* (London: Penguin Books, 2006), 274, 293.
14 For example, according to Wikipedia, 'Taipeil 101' (https://en.wikipedia.org/wiki/Taipei_101), it took five years to construct Taipei 101, recognised as the world's tallest building in 2004.
15 George Loewenstein et al., 'Introduction', in George Lowenstein et al., *Time and Decision: Economic and Psychological Perspectives on Intertemporal Choice* (New York: Russell Sage Foundation, 2003), 34. Also see Per Molander, *Condorcets misstag: hoten mot staten och demokratin* (Stockholm: Weyler, 2018), 192–93.
16 Ted O'Donoughe & Matthew Rabin, 'Self-Awareness and Self-Control', in Loewenstein et al., *Time and Decision*.
17 Daniel Kahneman, *Thinking Fast and Slow* (New York: Farrar, Straus & Giroux 2013).
18 See, for example, Drazen Prelec & Ronit Bodner, 'Self-Signaling and Self-Control', in Lowenstein et al., *Time and Decision*.
19 Loewenstein et al., *Time and Decision*, 10.
20 Ibid., 1–2.
21 Jean Piaget, *Six Psychological Studies* (New York: Vintage, Random House, 1968).
22 Sigmund Freud, *Introductory Lectures on Psycho-Analysis* (New York: Norton 1989).
23 See, for example, Jean Piaget, *The Child's Conception of the World* (Totowa: Rowman & Littlefield, 1989).
24 W. Mischel, *The Marshmallow Test: Mastering Self-Control* (New York: Little, Brown and Company, 2014), 17–18.
25 Ibid., 167.
26 Ibid., 167–68.
27 Ibid., 54–57, 72–74.
28 Ibid., 57.
29 Ibid., esp. ch. 7 and Part III in Mischel.
30 Ibid., 96–97 and passim; Walter Mischel et al., 'Sustaining Delay Gratification Over Time: A Hot–Cool System Perspective', in Loewenstein et al., *Time and Decision*, 179–80.
31 Tyler W. Watts et al., 'Revisiting the marshmallow Test: A Conceptual Replication Investigating Links between Early Delay of Gratification and Later Outcome', *Psychological Science* 29, no. 7, 2018, 1159–77.
32 Mischel, *The Marshmallow Test*, 124–27.
33 Ibid., 110.
34 Norbert Elias, *The Civilizing Process 1–2* (New York: Pantheon Books, 1982).

35 Hans Peter Duerr devoted five thick volumes to this attack on Elias, originally titled *Der Mythos vom Zivilisationsprozess*. I have read two of these tiresomely repetitive treatises, published in Swedish: *Nakenhet och skam (Volume 1)* (Stockholm: Symposion, 1994), and *Intimitet (Volume 2)* (Stockholm: Symposion, 1996); I have also read through a kind of pilot study for these volumes, his *Dreamtime* (Oxford: Basil Blackwell, 1985).

36 Arne Jarrick & Johan Söderberg, 'Inledning', in Arne Jarrick & Johan Söderberg (eds), *Människovärdet och makten: om civiliseringsprocessen i Stockholm 1600–1850* (Stockholm: Stockholmia förlag, 1994), 11–12.

37 Fabrizio Ricciardelli, 'Violence and Repression in Late Medieval Italy', in Samuel Kline Cohn Jr & Fabrizio Riccardielli (eds), *The Culture of Violence in Renaissance Italy* (Florence: Le Lettere, 2012), 56–59; Francesco Benigno, 'Reconsidering Popular Violence: Changes and Perspectives in the Analysis of Early Modern Revolts', in Samuel Kline Cohn Jr & Fabrizio Riccardielli (eds), *The Culture of Violence in Renaissance Italy*, 139; William Beik, 'The Violence of the French Crowd from Charivari to Revolution', *Past and Present*, no. 197, 2007, 86, 90 and n33, 100; Michel Mollat & Philippe Wolff, *The Popular Revolutions of the Middle Ages* (London: George Allen & Unwin, 1973), 51, 69, 73–74, 111; Vincent Challet, 'Violence as a Political Language: The Uses and Misuses of Violence in Late Medieval French and English Popular Rebellions', in Justine Finnhaber-Baker & Dirk Schoenaers, (eds), *The Routledge History Handbook of Medieval Revolt* (London: Routledge, 2017), 280; Paul Freedman, 'A Dossier of Peasant and Seigneurial Violence', in Justine Finnhaber-Baker & Dirk Schoenaers (eds), *The Routledge History Handbook of Medieval Revolt* (London: Routledge, 2017); J. K. Hyde, 'Contemporary Views on Factions and Civil Strife in Thirteenth- and Fourteenth-Century Italy', in Lauro Martines (ed.), *Violence and Civil Disorder in Italian Cities 1200–1500* (Berkeley: University of California Press, 1972), 289–90; Gene A. Brucker, 'The Florentine *popolo minuto* and Its Political Role', in Lauro Martines (ed.), *Violence and Civil Disorder in Italian Cities 1200–1500*, 179; Lauro Martines, 'Political Conflict in the Italian City States', *Government and Opposition* 3, no. 1, 1968, 70.

38 Manuel Eisner, 'From Swords to Words: Does Macro-Level Change in Self-Control Predict Long-Term Variation in Levels of Homicide?', *Crime and Justice* 43, no. 1, 2014, 1–72.

39 See Arne Jarrick & Johan Söderberg, *Praktisk historieteori* (Stockholm: Stockholms universitet, 2001), 125–26.

40 Jacques Le Goff, *The Birth of the Purgatory* (London: Scholar Press, 1984).

41 Jean Delumeau, *Le Péché et la peur: La culpabilisation en Occident [XIIIe-XVIIIe siecles]* (Paris: Fayard, 1983).

42 Fogel, *The Fourth Great Awakening*, ch. 5, esp. 178, 184, 190.

43 See, for example, John H. Goldthorpe et al., *The Affluent Worker in the Class Structure* (Cambridge: Cambridge University Press, 1969).

44 Leften S. Stavrianos, *Lifelines from Our Past: A New World History* (Armonk: M. E. Sharpe, 1997), 48, 53; Johan Goudsblom, *Fire and Civilization* (London: Allen Lane, 1992), 6–10; Stephen Mennell, 'Civilizing and Decivilizing Processes', in Johan Goudsblom et al. (eds), *The Course of Human History: Economic Growth, Social Process, and Civilization* (Armonk: M. E. Sharpe, 1996), 101; Johan Goudsblom, 'Human History and Long-Term Social Processes' in Johan Goudsblom et al. (eds), *The Course of Human History*, 22, 26; Franz Boas, *The Mind of Primitive Man* (New York: Macmillan, 1938), 132–33; Robert Carneiro, 'The Four Faces of Evolution', in John J. Honigman (ed.),

Handbook of Social and Cultural Anthropology (Chicago: Rand McNally, 1973), 95; Carroll Quigley, *The Evolution of Civilizations: An Introduction to Historical Analysis* (Indianapolis, Ind.: Liberty Fund, 1979), 211; John Robert McNeill & William McNeill, *The Human Web: A Bird's-Eye View of World History* (New York: W. W. Norton, 2003), 54.

45 See Eisner, 'From Swords to Words'.

Chapter 5. What Explains the Widespread Diffusion of Inequality and the Gradual Emergence of Egalitarianism Over the Centuries?

1   See Augusto López-Claros & Nakhjavani Bahiyyih, *Equality for Women = Prosperity for All* (New York: St Martin's Press, 2018).
2   J. Burns, *The Descent of Madness: Evolutionary Origins of Psychosis and the Social Brain* (London: Routledge, 2007), 74.
3   R. H. Tawney, *Equality* (London: George Allen & Unwin, 1964), 34.
4   For empirical support of such observations, see Michael Marmot, *Status Syndrome: How Your Social Standing Directly Affects Your Health and Life Expectancy* (London: Bloomsbury, 2004); Daniel Dorling, *Injustice: Why Social Inequality Persists* (Bristol: Policy Press, 2010), ch. 7, n. 1, 360; Malavika Subramanyam et al., 'Relative Deprivation in Income and Self-Rated Health in the United States', *Social Science & Medicine* 69, no. 3, 2009, 327–34.
5   López-Claros & Nakhjavani, *Equality for Women = Prosperity for All*, ch. 1.
6   Ibid.
7   Per Molander, *The Anatomy of Inequality: Its Social and Economic Origins and Solutions* (Brooklyn: Melville House Books, 2017), 125.
8   Peter Turchin, *War and Peace and War: The Rise and Fall of Empires* (London: Penguin Books, 2006), 263.
9   Ibid., 264–65.
10 Dorling, *Injustice*, 138.
11 Molander, *The Anatomy of Inequality*, 125; Turchin, *War and Peace and War*, 270.
12 For a more contemporary empirical and down-to-earth take on the issue, see Jan Kristian Pareliussen et al., 'Income Inequality in the Nordics from an OECD Perspective', in Rolf Aaberge et al. (eds), *Increasing Income Inequality in the Nordics. Nordic Economic Policy Review* (Copenhagen: Nordic Council of Ministers, 2018), 25.
13 Ricardo Fernholz & Robert Fernholz, 'Instability and Concentration in the Distribution of Wealth', *Journal of Economic Dynamics & Control* 44, July 2014, 251–69.
14 See, for instance, David Christian, *Maps of Time: An Introduction to Big History* (Berkeley: University of California Press, 2004), 259–64.
15 Christopher Boehm, *Hierarchy in the Forest: The Evolution of Egalitarian Behavior* (Cambridge, MA: Harvard University Press, 1999), 3–4.
16 Boehm, *Hierarchy in the Forest*, 234.
17 Ibid., 104–5.
18 Ibid., for example, 181–86, 198–99.
19 Ibid., 31.
20 Ibid., 31–32, 62, 93.
21 Ibid., 4.
22 Ibid., 65, 236.

23 Simon Kuznets, 'Economic Growth and Income Inequality', *American Economic Review* 45, no.1, 1955, 1–28.
24 Guido Alfani & Matteo Di Tullio, *The Lion's Share: Inequality and the Rise of the Fiscal State in Preindustrial Europe* (Cambridge: Cambridge University Press, 2019). This book meticulously measured both the proportion of poor and rich by Gini index and by top and bottom deciles.
25 Ibid., 11.
26 Thomas Piketty, *Capital in the Twenty-First Century* (Cambridge, MA: Belknap Press of Harvard University Press, 2014).
27 Ibid., 15, ch. 5, 8–9 and, for example, 165, 171, 181, 196, 291–92, 299–300, 316–19, 323–24, 327. Compare Aaberge et al. (eds), *Increasing Income Inequality in the Nordics*.
28 Piketty, *Capital in the Twenty-First Century*, for example, 154; also see 340–41.
29 Compare with Molander, *The Anatomy of Inequality*, 126; Jesper Roine, *Thomas Pikettys KAPITALET i det tjugoförsta århundradet* (Stockholm: Volante, 2014), 30–32; Dorling, *Injustice*, 128, 139, 191, 223, 256–60, also ch. 7 where he claims that mental disorders are more frequent in unequal societies, more so today.
30 Branko Milanovic, *The Haves and Have-nots* (New York: Basic Books, 2016), 'Introduction'.
31 Ibid., 158–59.
32 Ibid., 100–107, 116. This is even more stressed in Branko Milanovic, *Global Inequality: A New Approach for the Age of Globalization* (Cambridge, MA: Harvard University Press, 2016), ch. 1, esp. 35.
33 This paragraph is based on ibid., ch. 2.
34 Milanovic, *The Haves and Have-nots*, 16, 55. For a similar way of reasoning, see Lars Calmfors, 'Därför bör inte inkomstskillnaderna få fortsätta att växa i Sverige', *Dagens Nyheter*, 2 May 2018.
35 Tawney, *Equality*, 113. Also see 52, 93, 99.
36 Milanovic, *The Haves and Have Nots*, 100. Compare with Milanovic, *Global Inequality*, 39, 31–132.
37 Milanovic, *Global Inequality*, 132.
38 Walter Scheidel, *The Great Leveler: Violence and the History of Inequality from the Stone Age to the Twenty- First Century* (Princeton, NJ: Princeton University Press, 2017).
39 Ibid., 40, 62.
40 Ibid., 191.
41 Alfani & Di Tullio, *The Lion's Share*.
42 Pareliussen et al., 'Income Inequality in the Nordics …', 19–25.
43 Rolf Aaberge et al., 'Accounting for Public In-Kind Transfers in Comparison of Income Inequality between the Nordic Countries', in Aaberge et al. (eds), *Increasing Income Inequality in the Nordics*. Until 1918, in Sweden the right to vote was dependent on income.
44 This was one of the major notions in Tawney, *Equality*, 43, 52, 79, 93, 99, 113, 130, 149, etc.
45 Pareliussen et al., 'Income Inequality in the Nordics …'.
46 Dorling, *Injustice*, 91.
47 Ibid., 46, 48, 79.
48 Sir Arthur Lovejoy, *The Great Chain of Being* (Cambridge, MA: Harvard University Press, 1964); Arne Jarrick, *The Need to Be Needed: An Essay on Humankind, Culture, and World History* (Saarbrücken: Lambert Academic Publishing, 2010), ch. 2; Alfani & De Tullio, *The Lion's Share*, 58.

49 Georges Duby, *The Three Orders: Feudal Society Imagined* (Chicago: University of Chicago Press, 1980), 13, 19, 343–45; Hilding Pleijel, *Hustavlans värld* (Stockholm: Verbum, 1970), 30–52; Roland Mousnier, *Peasant Uprisings in Seventeenth-Century France, Russia, and China* (New York: Harper & Row, 1970), 233–41; Torbjörn Lodén, 'Inledning', in Torbjörn Lodén, (ed.), *Konfucius samtal* (Stockholm: Appell Förlag, 2016), 13–27.
50 Duby, *The Three Orders*, 74, 79–80.
51 Íbid., 86–87.
52 Íbid., 94.
53 Íbid., 19, chs 8–9, ch. 25; Pleijel, *Hustavlans värld*, ch. 1.
54 Duby, *The Three Orders*, 13, 19, 346.
55 Compare with Barrington Moore Jr, *Injustice: The Social Bases of Obedience and Revolt* (London: Macmillan, 1978), chs 1–3; Alfani & Di Tullio, *The Lion's Share*, 14.
56 For example, Ronny Ambjörnsson, *Människors undran: Europas idéhistoria – Antiken* (Stockholm: Natur & Kultur, 1997), 196–203.
57 Émile Coornaert, *Le compagnonnages en France du moyen âge à nos jours* (Paris: Editions ouvrières, 1966).
58 Duby, *The Three Orders*, 99–100, 102–6, 125.
59 Ibid., 105, 128.
60 Ibid., 130–34.
61 Ibid., 343, 345, epilogue, 354–56.
62 Karl Jaspers, *Vom Ursprung und Ziel der Geschichte* (Frankfurt/M & Hamburg: Fischer Bücherei, 1955); Karen Armstrong, *The Great Transformation: The Beginning of Our Religious Traditions* (New York: Anchor Books, 2007); S. N. Eisenstadt, 'Introduction: The Axial Age breakthroughs—their characteristics and origins', in S. N. Eisenstadt (ed.), *The Origins and Diversity of Axial Age Civilizations* (New York: State University of New York Press, 1986); Björn Wittrock, 'The meaning of Axial Age', in Johann P. Arnasson et al. (eds), *Axial Civilizations and World History* (Leiden: Brill, 2005).
63 Ayelet Shachar, *The Birthright Lottery: Citizenship and Global Inequality* (Cambridge, MA: Harvard University Press, 2009), 87, 89–108.
64 Arne Jarrick & Maria Wallenberg Bondesson, *The Dynamics of Law-Making: A World History* (Stockholm: KVHAA, 2018), and references given there.
65 Ibid.
66 Cited in Milanovic, *Global Inequality*, 142.
67 Shachar, *The Birthright Lottery*, see, for instance, 24–27.
68 Ibid., 62–63.
69 Martin Hägglund, *This Life: Why Mortalilty Makes Us Free* (London: Profile Books, 2019).

Chapter 6. Why Do People Appropriate Aesthetic Experience (Both as Producers and Consumers of Cultural Manifestations), and What Are the Individual and Societal Functions of Such Experiences?

1 Roger Scruton, *Beauty: A Very Short Introduction* (Oxford: Oxford University Press, 2009), 27–28, 49, 55, 69. Also see Donald E. Brown, *Human Universals* (New York: McGraw-Hill, 1991), 140.
2 For example, Steven Leuthold, *Indigenous Aesthetics* (Austin: University of Texas Press, 1998), esp. ch. 1; Scruton, *Beauty*, 113–14.

3   Sigmund Freud, *On Sexuality* (Harmondsworth: Penguin Books, 1986), 272. Compare Peter Turchin, *War and Peace and War: The Rise and Fall of Empires* (London: Penguin Books, 2006), 133–35.
4   Pierre Bourdieu, 'La production de la croyance. Contribution à une economie des biens symboliques', *Actes de la recherché en science sociales* 13, no. 3, 1977, 3–43; also see Daniel Dorling, *Injustice: Why Social Inequality Persists* (Bristol: Policy Press, 2010), 253–56.
5   On this I basically agree with Sherri Irvin, 'The Pervasiveness of the Aesthetic in Ordinary Experience', *British Journal of Aesthetic* 48, no. 1, 2008, 29–44, in her criticism of John Dewey.
6   Jessica J. Lee, 'Home Life: Cultivating a Domestic Aesthetic', *Contemporary Aesthetics* 8, no. 5, 2010.
7   Gary Iseminger, 'Experiential Theories of Aesthetic Value', in Richard Shusterman & Adele Tomlin (eds), *Aesthetic Experience*, New York: Routledge, 2008); Thomas Leddy, *The Extraordinary in the Ordinary* (Ontario: Broadview Press, 2012), 38, 64. Also see Yuriko Saito, *Everyday Aesthetics* (Oxford: Oxford University Press, 2007), 10; Robert Stecker, 'Plato', in Alessandro Giovannelli (ed.), *Aesthetics: The Key Thinkers* (London: Continuum International, 2012), 13.
8   Elisabeth Schellekens, 'Immanuel Kant', in Giovannelli (ed.), *Aesthetics*.
9   Compare with Leddy, *The Extraordinary in the Ordinary*, 22–23.
10 Ibid., 72.
11 Ibid. Also see Saito, *Everyday Aesthetics*, 25–26, though Saito thinks that artists fails to capture the "everyday-ness" of everyday life by sticking to the distinction.
12 Leddy, *The Extraordinary in the Ordinary*, 13.
13 See, for example, Saito, *Everyday Aesthetics*, 55; Peter Bryngelsson, *Musik på blodigt allvar: en studie av musikens roll i krig och konflikter* (Möklinta: Gidlunds förlag, 2017).
14 See, for example, Leddy *The Extraordinary in the Ordinary*, 46–47; Saito, *Everyday Aesthetics*, 25–28.
15 Leddy, *The Extraordinary in the Ordinary*, 46–47; see Scruton, *Beauty*, 69–71.
16 John Dewey, *Art as Experience* (New York: Penguin Books, 1980), 41.
17 Gian Carlo Garfagnini, 'Medieval Aesthetics', in Giovannelli (ed.), *Aesthetics*, 34, 43, 46.
18 Ibid., 40–41.
19 Compare Sven Ove Hansson, 'Aesthetic Functionalism', *Contemporary Aesthetic 3*, 2005, https://contempaesthetics.org/newvolume/pages/article.php?articleID=324; Leddy, *The Extraordinary in the Ordinary*, 26, 66; Scruton, *Beauty*, 15–18, 66.
20 Saito, *Everyday Aesthetics*, 26.
21 Compare with Immanuel Kant, *Kritik av omdömeskraften* (Stockholm: Thales, 2003), 57, 66, 69; also see Immanuel Kant, 'A Theory of Aesthetic Judgment: From *The Critique of Judgment*', in George Dickie & Richard J. Sclafani (eds), *Aesthetics: A Critical Anthology* (New York: St Martin's Press, 1977).
22 See Garfagnini, 'Medieval Aesthetics', 45, for a comparison with William of Okham's views.
23 Dewey, *Art as Experience*, 6, 31.
24 Compare Scruton, *Beauty*, xi.
25 Ibid., 76.
26 Ibid., 8–10, where this is called *aesthetic minimalism*.
27 Saito, *Everyday Aesthetics*, 22, 46.
28 Ibid., 120, and the following pages.

29 Peter Bogucki, *The Origins of Human Society* (Malden, MA: Blackwell Publishers, 1999), 43–44; Robert Boyd & Joan B. Silk, *How Humans Evolved* (New York: W. W. Norton, 2000), 418.
30 Compare with Scruton, *Beauty*, ch. 3, for example, 50.
31 Ibid., 73.
32 Saito, *Everyday Aesthetics*, 122. Compare with Scruton, *Beauty*, 1.
33 Susan Feagin, 'Roger Fry [1866–1934] and Clive Bell [1881–1964]', in Giovannelli (ed.), *Aesthetics*, 116.
34 Leddy, *The Extraordinary in the Ordinary*, 19, 45. Also see p. 20; Saito, *Everyday Aesthetics*, 5, 11.
35 For rich overviews, see Dicke & Sclafani (eds), *Aesthetics*; Giovanelli (ed.), *Aesthetics*.
36 Steven Mithen, *The Singing Neanderthals. The Origin of Music* (London: Weidenfeld & Nicholson, 2005); Steven Pinker, *How the Mind Works* (New York: W. W. Norton, 1997). Both are referred to in Jan Fagius, 'Musik – ett evolutionens barn?', in Jenny Björkman & Arne Jarrick (eds), *RJ:s årsbok 2018* (Göteborg & Stockholm: Makadam förlag, 2018); Lisa Zunshine, *Why We Read Fiction: Theory of the Mind and the Novel* (Columbus: Ohio State University Press, 2006).
37 Dewey, *Art as Experience*; Thomas Leddy, 'John Dewey', in Giovannelli (ed.), *Aesthetics*.
38 Some scattered examples are Saito, *Everyday Aesthetics*, 58, 61; Kevine Melchionne, 'The Definition of Everyday Aesthetics', *Contemporary Aesthetics*, 51, no. 4, 2011, 441. Also see Dicke & Sclafani (eds), *Aesthetics*, containing most of the famous ideas on the issue but being dry in terms of data.
39 Scruton, *Beauty*, passim, esp. 124, 145.
40 Ibid., 71; Leddy, *The Extraordinary in the Ordinary*, 17; Saito, *Everyday Aesthetics*.
41 Dewey, *Art as Experience*, 48.
42 Leddy, *The Extraordinary in the Ordinary*, 44. Also see 77.
43 Dewey, *Art as Experience*, 3, 11. For a similar view on Dewey, see Leddy, *The Extraordinary in the Ordinary*, 44. Also see Leddy, 'John Dewey', 126. Another typical example is C. McManus & A. Furham, 'Aesthetic Activities: Influences of Education, Background and Personality on Interest and Involvement in the Arts', *British Journal of Psychology* 97, no. 4, 2006, 555–87; Edward Warder Rannells, 'Aesthetic Expression and Learning', *Journal of Aesthetics and Art Criticism*, 5, no. 4, 1947, 314–20.
44 Some of the more influential are Melchionne, 'The Definition of Everyday Aesthetics'; Saito, *Everyday Aesthetics*; Irvin, 'The Pervasiveness of the Aesthetic in Ordinary Experience', *British Journal of Aesthetics* 48, no. 1, 2008, 29–44; Leddy, *The Extraordinary in the Ordinary*.
45 Melchionne, 'The Definition of Everyday Aesthetics'.
46 For one such example, see Alessandro Giovanelli, 'Some Contemporary Developments', in Giovanelli (ed.), *Aesthetics*.
47 Melchionne, 'The Definition of Everyday Aesthetics'; Kevin Melchionne, 'Aesthetic Experience in Everyday Life: A Reply to Dowling', *British Journal of Aesthetics* 51, no. 4, 2011, 437–42; Christopher Dowling, 'The Aesthetics of Daily Life', *British Journal of Aesthetics* 51, no 3, 2010, 225–42.
48 Compare Leddy, *The Extraordinary in the Ordinary*, 46; Saito, *Everyday Aesthetics*.
49 See Irvin, 'The Pervasiveness of the Aesthetic in Ordinary Experience'; Dewey, *Art as Experience*, ch. 2; Leddy, *The Extraordinary in the Ordinary*, 58, 76–77.
50 Compare with Saito, *Everyday Aesthetics*, 153.
51 See, for example, Shusterman & Tomlin (eds), *Aesthetic Experience*.

52 Compare with Paul Duncan, quoted in Saito, *Everyday Aesthetics*, 14.
53 This is the essential idea in Charles Darwin's, *The Descent of Man and Selection in Relation to Sex* (Amherst, NY: Prometheus Books, 1997).
54 Scruton, *Beauty*, 31, 32.
55 Ibid., 35, 43; Plato, 'The Symposium', in *The Portable Plato* (Harmondsworth: Penguin Books, 1976); Gerhard Richter, 'Walter Benjamin [1892–1940] and Theodor Adorno [1903–1969]', in Giovannelli (ed.), *Aesthetics*, 150.
56 See, for example, Feagin, 'Roger Fry …', 114.
57 Compare with Scruton, *Beauty*, 145–47, 156.

III. What Lies Ahead?

1 For one influential example of this, see Robert Boyd & Peter J. Richerson, *Culture and the Evolutionary Process* (Chicago: University of Chicago Press, 1985). In 2007, The Centre for the Study of Cultural Evolution was inaugurated at Stockholm University, a genuinely interdisciplinary structure comprising researchers in biology, mathematics, history, archaeology, linguistics and so on.
2 This is an intermittently recurring concern. See, for instance, Thord Silverbark, *Fysikens filosofi: Diskussioner om Einstein, relativitetsteorin och kvantfysiken i Sverige 1910–1970* (Stockholm: Brutus Östlings förlag, 1999), 90–91, 106.
3 For the conventional perspective, see Peter Burke *What Is the History of Knowledge?* (Cambridge: Polity Press, 2016), 64–66.
4 Compare ibid., 6–7.
5 For a particularist view, ibid., 123, although contradicted (unconsciously?) on 125.
6 Arne Jarrick & Maria Wallenberg Bondesson, *The Dynamics of Law-Making: A World History* (Stockholm: KVHAA, 2018), 34–39.
7 Compare with Burke, *What Is the History of Knowledge?*, 60–61.

A. Five Thematic Clusters Summarising a Workshop on Big Questions

1 Based on Lena Johansson de Chateau's minutes from the workshop.

B. Translated Highlights from an Article on the Big Research Questions

1 Arne Jarrick & Janken Myrdal, 'De stora forskningsfrågorna om människans villkor', *Respons*, no. 5, 2015.

REFERENCES

Aaberge, Rolf, et al. (eds). *Increasing Income Inequality in the Nordics. Nordic Economic Policy Review*, Copenhagen: Nordic Council of Ministers, 2018.

Aaberge, Rolf, et al. 'Accounting for Public In-Kind Transfers in Comparison of Income Inequality between the Nordic Countries'. In *Increasing Income Inequality in the Nordics. Nordic Economic Policy Review 2018*, edited by Rolf Aaberge et al., 175–205. Copenhagen: Nordic Council of Ministers, 2018..

Åberg, Rune. 'Unemployment Persistency. Over-Education and the Employment of the Less Educated'. *European Sociological Review* 19, no. 2 (2003): 199–216.

Alfani, Guido, & Matteo, Di Tullio. *The Lion's Share: Inequality and the Rise of the Fiscal State in Preindustrial Europe*. Cambridge: Cambridge University Press, 2019.

Allcott, Hunt, & Gentzkow, Matthew. 'Social Media and Fake News in the 2016 Election'. *Journal of Economic Perspectives* 31, no. 2 (2017): 211–36.

Åmark, Klas. *Att bo granne med ondskan: Sveriges förhållande till nazismen, Nazityskland och förintelsen*. Stockholm: Bonniers, 2011.

Ambjörnsson, Ronny. *Människors undran (Europas idéhistoria: Antiken)*. Stockholm: Natur & Kultur, 1997.

Arendt, Hannah. *Eichmann in Jerusalem: A Report on the Banality of Evil*. New York: Viking Press, 1963.

Armstrong, Karen. *The Great Transformation: The Beginning of Our Religious Traditions*. New York: Anchor Books, 2007.

Asch, S. E. *Social Psychology*. New York: Prentice-Hall, 1952.

Beik, William. 'The Violence of the French Crowd from Charivari to Revolution'. *Past and Present*, no. 197 (2007): 75–110.

Bell, Daniel. *The Coming of the Post-Industrial Society: A Venture in Social Forecasting*. New York: Basic Books, 1973.

Benigno, Francesco. 'Reconsidering Popular Violence: Changes and Perspectives in the Analysis of Early Modern Revolts. In *The Culture of Violence in Renaissance Italy*, edited by Samuel Kline Cohn Jr & Fabrizio Riccardielli, 123–44. Florence: Le Lettere, 2012.

Blass, Thomas (ed.). *Obedience to Authority: Current Perspectives on the Milgram Paradigm*. Mahwah, NJ: Lawrence Erlbaum Associate, 2000.

———. 'The Milgram Paradigm after 35 Years: Some Things We Now Know about Obedience to Authority'. In *Obedience to Authority: Current Perspectives on the Milgram Paradigm*, edited by Thomas Blass, 955–78. Mahwah, NJ: Lawrence Erlbaum Associate, 2000.

Boas, Franz. *The Mind of Primitive Man*. New York: Macmillan, 1938.

Boehm, Christopher. *Hierarchy in the Forest: The Evolution of Egalitarian Behavior*. Cambridge, MA: Harvard University Press, 1999.

Bogucki, Peter. *The Origins of Human Society*. Malden, MA: Blackwell, 1999.

Bond, Rod, & Smith, Peter J. 'Culture and Conformity: A Meta-Analysis of Studies Using Asch's (1952b, 1956) Line Judgment Task'. *Psychological Bulletin* 119, no. 1 (1996): 111–37.
Bourdieu, Pierre. 'La production de la croyance. Contribution à une economie des biens symboliques'. *Actes de la recherché en science sociales* 13, no. 3 (1977): 3–43.
Boyd, Robert, & Richerson, Peter J. *Culture and the Evolutionary Process*. Chicago: University of Chicago Press, 1985.
Boyd, Robert, & Silk, Joan B. *How Humans Evolved*. New York: W. W. Norton, 2000.
Braudel, Fernand. 'History and the Social Sciences: The *Longue Durée*'. In *On History*, edited by Fernand Braudel, 25–54. Chicago: University of Chicago Press, 1980.
———. *The Mediterranean and the World in the Age of Philip II*. London: Collins, 1972–73.
Bremmer, Magnus. 'Den frånvarande humanisten kan vitalisera vårt fantasilösa idéklimat'. *Dagens Nyheter*, 15 March 2019.
Browning, Christopher. *Ordinary Men: Reserve Police Battalion 101 and the Final Solution in Poland*. New York: HarperCollins, 1992.
Brucker, Gene A. 'The Florentine *Popolo Minuto* and Its Political Role. In *Violence and Civil Disorder in Italian cities 1200–1500*, edited by Lauro Martines, 155–83. Berkeley: University of California Press, 1972.
Bryngelsson, Peter. *Musik på blodigt allvar: en studie av musikens roll i krig och konflikter*. Möklinta: Gidlunds förlag, 2017.
Burke, Peter. *What Is the History of Knowledge?* Cambridge: Polity Press, 2016.
Burns, Jonathan. *The Descent of Madness: Evolutionary Origins of Psychosis and the Social Brain*. London: Routledge, 2007.
Calmfors, Lars. 'Därför bör inte inkomstskillnaderna få fortsätta att växa i Sverige'. *Dagens Nyheter*, 05 February 2018.
———. 'Migration, sysselsättning och löner. Hur ska vi ta itu med arbetsmarknadens tudelning. In *Immigration, ekonomisk integration och entreprenörskap*, edited by Johan Lundberg, 18–37. Stockholm: Swedish Economic Forum, 2016.
Carneiro, Robert L. 'Four Faces of Evolution'. In *Handbook of Social and Cultural Anthropology*, edited by John J. Honigman, 89–109. Chicago: Rand McNally, 1973.
Centola, Damon, et al. 'Experimental Evidence for Tipping Points in Social Convention'. *Science* 360, no. 6393 (2018): 1116–19.
Challet, Vincent. 'Violence as a Political Language: The Uses and Misuses of Violence in Late Medieval French and English Popular Rebellions'. In *The Routledge History Handbook of Medieval Revolt*, edited by Juistine Finnhaber-Baker & Dirk Schoenaers, 279–91. London: Routledge, 2017.
Chartier, Roger. *Cultural History: Between Practices and Representations*. Cambridge: Polity in association with Blackwell, 1988.
Christian, David. *Maps of Time: An Introduction to Big History*. Berkeley: University of California Press, 2004.
Cinnirella, Marco, & Green, Ben. 'Does "Cyber-Conformity" Vary Cross-Culturally? Exploring the Effect of Culture and Communication Medium on Social Conformity'. *Computers in Human Behavior* 23, no. 4 (2007): 2011–25.
Cohn Jr, Samuel Kline, & Riccardielli, Fabrizio (eds). *The Culture of Violence in Renaissance Italy*. Florence: Le Lettere, 2012.
Comte, Auguste. *A General View of Positivism*. London: Trübner, 1865.
Cook, John, & Lewandowsky, Stephan. 'The Debunking Handbook'. 27 November 2011. https://skepticalscience.com/Debunking-Handbook-now-freely-available-download.html.

Coornaert, Émile. *Le compagnonnages en France du moyen âge à nos jours*. Paris: Editions ouvrières, 1966.
Cutler, Stephen J., & Kaufman, Robert L. 'Cohort Changes in Political Attitudes: Tolerance of Ideological Nonconformity'. *Public Opinion Quarterly* 39, no.1 (1975): 69–81.
Darwin, Charles. *The Descent of Man and Selection in Relation to Sex*. Amherst, NY: Prometheus Books, 1997.
———. *Självbiografi*. Stockholm: Natur & kultur, 2009.
Daum, Andreas W. 'Varieties of Popular Science and the Transformation of Public Knowledge: Some Historical Reflections'. *Isis* 100, no. 2 (2009): 319–32.
Delumeau, Jean. *Le Péché et la peur: La culpabilisation en Occident (XIIIe-XVIIIe siecles)*. Paris: Fayard, 1983.
Dewey, John. *Art as Experience*. New York: Penguin, 1980.
Dickie, George, & Sclafani, Richard J. (eds). *Aesthetics: A Critical Anthology*. New York: St Martin's Press, 1977.
Dorling, Daniel. *Injustice: Why Social Inequality Persists*. Bristol: Policy Press, 2010.
Dowling, Christopher. 'The Aesthetics of Daily Life'. *British Journal of Aesthetics* 51, no. 3 (2010): 225–42.
Drucker, Peter F. *The Age of Discontinuity. Guidelines to Our Changing Society*. London: Harper & Row, 1969.
Duby, Georges. *The Three Orders: Feudal Society Imagined*. Chicago: University of Chicago Press, 1980.
Duerr, Hans Peter. *Dreamtime*. Oxford: Basil Blackwell, 1985.
———. *Nakenhet och skam (volume 1)*. Stockholm: Symposion, 1994.
———. *Intimitet (volume 2)*. Stockholm: Symposion, 1996.
Edwards, K., & Smith, E. E. 'A Disconfirmation Bias in the Evaluation of Arguments'. *Journal of Personality and Social Psychology*, 71, no. 1 (1996): 5–24.
Efron, Noah J. 'Myth 9. That Christianity Gave Birth to Modern Science'. In *Galileo Goes to Jail and Other Myths about Science and Religion*, edited by Ronald L. Numbers, 79–89. Cambridge, MA: Harvard University Press, 2009.
Eisenstadt, S. N. 'Introduction: The Axial Age Breakthroughs–Their Characteristics and Origins'. In *The Origins and Diversity of Axial Age Civilizations*, edited by S. N. Eisenstadt, 1–25. New York: State University of New York Press, 1986.
Eisner, Manuel. 'From Swords to Words: Does Macro-Level Change in Self-Control Predict Long-Term Variation in Levels of Homicide?' *Crime and Justice* 43, no. 1 (2014): 1–72.
Ekström, Anders (ed.). *Den mediala vetenskapen*. Nora: Nya Doxa, 2004.
Elias, Norbert. *The Civilizing Process 1–2*. New York: Pantheon Books 1982.
Ericsson, K. A. 'An Introduction to the Cambridge Handbook of Expertise and Expert Performance: Its Development, Organization, and Content'. In *The Cambridge Handbook of Expertise and Expert Performance*, edited by K. A. Ericsson et al., 3–20. Cambridge: Cambridge University Press, 2006.
Eriksson, Robert. 'Why Do Graduates Live Longer?' In *Cradle to Grave: Life-Course in Modern Sweden*, edited by J. O. Jonsson & C. Mills. Durham, NC: Sociology Press, 2001.
Fagius, Jan. 'Musik – ett evolutionens barn?' In *RJ:s årsbok 2018*, edited by Jenny Björkman & Arne Jarrick, 17–32. Göteborg & Stockholm: Makadam förlag, 2018.
Feagin, Susan. 'Roger Fry (1866–1934) and Clive Bell (1881–1964)'. In *Aesthetics: The Key Thinkers*, edited by Alessandro Giovannelli, 113–25. London: Continuum International, 2012.

Fernholz, Ricardo, & Fernholz, Robert. 'Instability and Concentration in the Distribution of Wealth'. *Journal of Economic Dynamics & Control* 44, July (2014): 251–69.
Festinger, Leon. *A Theory Cognitive Dissonance*. Stanford, CA: Stanford University Press, 1957.
Finnhaber-Baker, Justine, & Schoenaers, Dirk (eds). *The Routledge History Handbook of Medieval Revolt*. London: Routledge, 2017.
Fitzharris, Lindsey. *The Butchering Art: Joseph Lister's Quest to Transform the Grisly World of Victorian Medicine*. New York: Scientific American/Farrar Straus and Giroux, 2017.
Flynn, J. 'Massive IQ Gains in 14 Nations: What IQ Tests Really Measure'. *Psychological Bulletin* 101, no. 2 (1987): 171–91.
Fogel, Robert. *The Fourth Great Awakening & the Future of Egalitarianism*. Chicago: University of Chicago Press, 2000.
Forskningsöversikt 2019: Humaniora och samhällsvetenskap . Stockholm: Vetenskapsrådet, 2019.
Freud, Sigmund. *Introductory Lectures on Psycho-Analysis*. New York: Norton, 1989.
———. *On Sexuality*. Harmondsworth: Penguin Books, 1986.
Galtung, Johan, & Inayatullah, Sohail. *Macrohistory and Macrohistorians: Perspectives on Individual, Social, and Civilizational Change*. Westport, CT: Praeger, 1997.
Garfagnini, Gian Carlo. 'Medieval Aesthetics'. In *Aesthetics: The Key Thinkers*, edited by Alessandro Giovannelli, 34–47. London: Continuum International, 2012.
Ghirlanda, Stefano, et al. 'Memory for Stimulus Sequences: A Divide between Humans and Other Animals?'. *Royal Society Open Science* (21 June 2017). https://royalsocietypublishing.org/doi/full/10.1098/rsos.161011.
Gilens, Martin. 'Political Ignorance and Collective Policy Preferences'. *American Political Science Review* 95, no. 2 (2001): 379–96.
Ginzburg, Carlo. *The Cheese and the Worms: The Cosmos of a Sixteenth-Century Miller*. London: Routledge, 1980.
Giovanelli, Alessandro. 'Some Contemporary Developments. In *Aesthetics: The Key Thinkers*, edited by Alessandro Giovannelli, 207–30. London: Continuum International, 2012.
Giovannelli, Alessandro (ed.). *Aesthetics: The Key Thinkers*. London: Continuum International, 2012.
Girskevicius, Vladas, et al. 'Going Along versus Going Alone: When Fundamental Motives Facilitate Strategic (Non)Conformity'. *Journal of Personality and Social Psychology* 91, no. 2 (2006): 281–94.
Goldthorpe, John H., et al. *The Affluent Worker in the Class Structure*. Cambridge: Cambridge University Press, 1969.
Gredebäck, Gustaf, & Annika, Melinder. 'Infants' Understanding of Everyday Social Interactions: A Dual Process Account'. *Cognition* 114, no. 2 (2010): 197–206.
Goudsblom, Johan. *Fire and Civilization*. London: Allen Lane, 1992.
———. 'Human History and Long-Term Social Processes'. In *The Course of Human History: Economic Growth, Social Process, and Civilization*, edited by Johan Goudsblom et al., 15–30. New York: Routledge, 1996.
Guilbeault, Douglas, et al. 'Social Learning and Partisan Bias in the Interpretation of Climate Trends'. *PNAS* 115 (2018): 9714–19.
Guille-Escuret, Georges. 'Technical Innovation and Cultural Resistance'. In *Technological Choices: Transformation in Material Cultures since the Neolithic*, edited by Pierre Lemonnier, 214–26. London: Routledge, 1993.
Hägglund, Martin. *This Life: Why Mortality Makes Us Free*. London: Profile Books, 2019.
Hallberg, Anna Victoria. *Hotet mot humaniora kommer inifrån*. Stockholm: Timbro, 2019.

Hammar, Isak. 'Den klassiska folkbildaren kanske tillhör historien'. *Dagens Nyheter*, 22 March 2019.
Haney, Graig, et al. 'Interpersonal Dynamics in a Simulated Prison'. *International Journal of Criminology and Penology* 1, no. 1 (1973): 69–97.
Hansson, Sven Ove. 'Aesthetic Functionalism'. *Contemporary Aesthetic* 3 (2005). https://contempaesthetics.org/newvolume/pages/article.php?articleID=324.
Hart, H. L. A. *The Concept of Law*. Oxford: Oxford University Press, 1997.
Haslam, Alexander S., & Reicher, Stephen D. 'The Psychology of Tyranny'. *Scientific American Mind* 16, no. 3 (2005): 44–51.
Haslam, S. Alexander, & Reicher, Stephen. 'Beyond the Banality of Evil: Three Dynamics of an Interactionist Social Psychology of Tyranny'. *Personality and Social Psychology Bulletin* 33, no. 5 (2007): 615–22.
Herlihy, David V. *Bicycle: The History*. New Haven, CT: Yale University Press, 2004.
Hirschman, Albert O. *Exit, Voice and Loyalty: Responses to Decline in Firms, Organizations, and States*. Cambridge, MA: Harvard University Press, 1970.
———. *Shifting Involvements: Private Interest and Public Action*. Princeton, NJ: Princeton University Press, 2002.
Hodges, Bert H., & Geyer, Anne L. 'A Nonconformist Account of the Asch Experiments: Values, Pragmatics, and Moral Dilemmas'. *Personality and Social Psychology Review* 10, no. 1 (2006): 2–19.
Hofstede, Geert. 'The Cultural Relativity of Organizational Practices and Theories'. *Journal of International Business Studies* 14, no. 2(1983): 75–89.
Hollsten, Laura. 'Public, Private, and Experience-Based Knowledge: Cholesterol Knowledge in Circulation in Finnish Society, 1970–2010'. In *Circulation of Knowledge: Explorations in the History of Knowledge*, edited by Johan Östling et al, 37–55. Lund: Nordic Academic Press, 2018.
Holm, Poul Holm, et al. *The Humanities World Report*. Basingstoke: Palgrave Macmillan, 2015.
Hornsey, Matthew J. 'Ingroup Critics and Their Influence on Groups'. In *Individuality and the Group: Advances in Social Identity*, edited by Tom Postmes & Jolanda Jetten, 74–92. London: Sage 2006.
Hornsey, Matthew J., et al. 'On Being Loud and Proud: Non-Conformity and Counter-Conformity to Group Norms'. *British Journal of Social Psychology* 42, no. 3 (2003): 319–35.
Hyde, J. K. 'Contemporary Views on Factions and Civil Strife in Thirteenth- and Fourteenth-Century Italy'. In *Violence and Civil Disorder in Italian Cities 1200–1500*, edited by Lauro Martines, 273–307. Berkeley: University of California Press, 1972.
IAOS. '80 Former Chief Statisticians Condemn Prosecution of Andreas Georgiou'. 18 June 2018. https://www.iaos-isi.org/index.php/latestnews/221-80-former-chief-statisticians-condemn-prosecution-of-andreas-georgiou.
Ibsen, Hilde. 'Global økologisk historie'. In *Jorden runt igen – nya bidrag till en gammal globalhistoria* (25:e Nordiska historiemötet 2004), edited by Arne Jarrick & Alf Johansson, 67–85. Stockholm: Almqvist & Wiksell International, 2004.
Ifrah, Georges. *The Universal History of Numbers: From Prehistory to the Invention of the Computer, Part 1*. London: Harvill Press, 1998.
Irvin, Sherri. 'The Pervasiveness of the Aesthetic in Ordinary Experience'. *British Journal of Aesthetic* 48, no. 1 (2008): 29–44.
Iseminger, Gary. 'Experiential Theories of Aesthetic Value'. In *Aesthetic Experience*, edited by Richard Shusterman & Adele Tomlin, 45–58. New York: Routledge, 2008.

Isidorsson, Tommy. *Striden om tiden: Arbetstidens utveckling i Sverige under 100 år i ett internationellt perspektiv*. Göteborg: Department of History, 2001.

Izuma, Keise. 'The Neural Basis of Social Influence and Attitude Change'. *Current Opinion in Neurobiology* 23, no. 3 (2013): 456–62.

Jami, Catherine. 'The Jesuits' Negotiation of Science between France and China (1685–1722): Knowledge and Modes of Imperial Expansion'. In *Negotiating Knowledge in Early-Modern Empires: A Decentered View*, edited by Lázló Kontler et al., 53–77. New York: Palgrave Macmillan, 2014.

Jansson, Fredrik, et al. 'Democratic Revolutions as Institutional Innovation Diffusion: Rapid Adoption and Survival of Democracy'. *Technological Forecasting & Social Change* 80, no. 8 (2013): 1546–56.

Jardine, Nicholas, & Emma, Spary. 'Worlds of History'. In *Worlds of Natural History*, edited by H. A. Curry et al., 3–14. Cambridge: Cambridge University Press, 2018.

Jarrick, Arne. *Back to Modern Reason: Johan Hjerpe and Other Petit Bourgeois in Stockholm in the Age of Enlightenment*. Liverpool: Liverpool University Press, 1999.

———. *The Need to Be Needed: An Essay on Humankind, Culture, and World History*. Saarbrücken: Lambert Academic, 2010.

———. *Översyn av humaniora vid Örebro universitet*. Örebro: Örebro universitet, 2015.

———. 'The Pressure to Conform, the Need to Rebel: A Historical Project on Resisting Group Pressure'. In *Långa linjer och många fält. Festskrift till Johan Söderberg*, edited by Martin Gustavsson & Dag Retsö, 231–56. Stockholm: Stockholms universitet, 2015.

———. *Det finns inga häxor: en bok om kunskap*. Stockholm: Weyler, 2017.

———. 'Knowledge Resistance: A Global Challenge – in Research and Education, in the Humanities and Elsewhere'. In *Humanities and Higher Education: Synergies between Science, Technology and Humanities (Higher Education in the World 7)*, edited by David Bueno et al., 77–84. Barcelona: GUNi network, 2019.

Jarrick, Arne, & Myrdal, Janken. 'De stora forskningsfrågorna om människans villkor'. *Respons*, no. 5 (2015): 46–49.

Jarrick, Arne, et al. (eds). *Methods in World History: A Critical Approach*. Lund: Nordic Academic Press, 2016.

Jarrick, Arne, & Pagin, Peter. 'Democracy and Academic Freedom – the Hungarian Example'. 10 December 2019. https://kva.se/en/nyheter/demokrati-och-akademisk-frihet-exemplet-ungern.

Jarrick, Arne, & Söderberg, Johan. 'Inledning'. In *Människovärdet och makten: om civiliseringsprocessen i Stockholm 1600–1850*, edited by Arne Jarrick & Johan Söderberg, 9–20. Stockholm: Stockholmia förlag, 1994.

———. *Praktisk historieteori*. Stockholm: Stockholms universitet, 2001.

Jarrick, Arne, & Wallenberg Bondesson, Maria. 'Flexible Comparativeness: Towards Better Methods for the Cultural Historical Study of Laws and Other Aspects of Human Culture'. In *Organizing History: Studies in Honour of Jan Glete*, edited by Anna Maria Forssberg et al., 179–99. Lund: Nordic Academic Press, 2011.

———. *The Dynamics of Law-Making: A World History*. Stockholm: KVHAA, 2018.

Jaspers, Karl. *Vom Ursprung und Ziel der Geschichte*. Frankfurt: Fischer Bücherei, 1955.

Johansson, Alf. *Den effektiva arbetstiden: Verkstäderna och arbetsintensitetens problem 1900–1920*. Uppsala: Uppsala University, 1977.

Jonsson, Kjell, & Larsson, Jenny. 'Mellan vetenskap och vers: Ärftlighetsforskare som publika vetenskapsmän 1940–1960'. In *Den mediala vetenskapen*, edited by Anders Ekström, 217–49. Nora: Nya Doxa, 2004.

Kahan, Dan, et al. 'Motivated Numeracy and Enlightened Self-Government'. *Behavioural Public Policy* 1, no. 1 (2017): 54–86.
Kahneman, Daniel. *Thinking Fast and Slow*. New York: Farar, Straus & Giroux, 2013.
Kant, Immanuel. 'A Theory of Aesthetic Judgment: From *The Critique of Judgment*'. In *Aesthetics: A Critical Anthology*, edited by George Dickie & Richard J. Sclafani, 287–305. New York: St Martin's Press, 1977.
———. *Kritik av omdömeskraften*. Stockholm: Thales, 2003.
Keynes, John M. *The General Theory of Employment, Interest and Money*. London: Macmillan, 1973.
Kuznets, Simon. 'Economic Growth and Income Inequality'. *American Economic Review* 45, no.1 (1955): 1–28.
Lane, Robert. 'The Decline of Politics and Ideology in a Knowledgeable Society'. *American Sociological Review* 31, no. 5 (1965): 649–42.
Larsson, Göran. *Humaniora – en vetenskap bland andra*. Stockholm: Timbro, 2019.
Lasch, Christopher. *The Culture of Narcissism: American Life in an Age of Diminishing Expectations*. New York: W.W. Norton, 1978.
Le Goff, Jacques. *The Birth of the Purgatory*. London: Scholar Press, 1984.
Leddy, Thomas. *The Extraordinary in the Ordinary*. Ontario: Broadview Press, 2012.
———. 'John Dewey'. In *Aesthetics: The Key Thinkers*, edited by Alessandro Giovannelli, 126–38. London: Continuum International, 2012.
Lee, Jessica J. 'Home Life: Cultivating a Domestic Aesthetic'. *Contemporary Aesthetics* 8, no. 5 (2010). http://hdl.handle.net/2027/spo.7523862.0008.015.
Lemonnier, Pierre. 'Introduction'. In *Technological Choices: Transformation in Material Cultures since the Neolithic*, edited by Pierre Lemonnier, 1–35. London: Routledge, 1993.
Le Roy Ladurie, Emmanuel. *The Peasants of Languedoc*. Urbana: University of Illinois, 1974.
Leuthold, Steven. *Indigenous Aesthetics*. Austin: University of Texas Press, 1998.
Lindenfors, Patrik, et al. 'The Cultural Evolution of Democracy: Saltational Changes in a Political Regime Landscape'. *PLoS One* 6 (30 November 2011): 1–4, https://journals.plos.org/plosone/article?id=10.1371/journal.pone.0028270.
Lindroth, Sten. *Svensk lärdomshistoria Gustavianska tiden*. Stockholm: Norstedts, 1989.
Lodén, Torbjörn. 'Inledning'. In *Konfucius samtal*, edited by Torbjörn Lodén, 7–28. Stockholm: Appell Förlag, 2016.
Loewenstein, George, et al. 'Introduction'. In *Time and Decision: Economic and Psychological Perspectives on Intertemporal Choice*, edited by George Lowenstein et al., 1–12. New York: Russell Sage Foundation, 2003.
Logan, Michael H., & Qirko, Hector N. 'An Evolutionary Perspective on Maladaptive Traits and Cultural Conformity'. *American Journal of Human Biology* 8, no. 5 (1996): 615–29.
López-Claros, Augusto, & Bahiyyih, Nakhjavani. *Equality for Women = Prosperity for All*. New York: St Martin's Press, 2018.
Lord, Charles G., et al. 'Biased Assimilation and Attitude Polarization: The Effects of Prior Theories on Subsequently Considered Evidence'. *Journal of Personality and Social Psychology* 37, no. 11 (1979): 2098–109.
Lovejoy, Sir Arthur. *The Great Chain of Being*. Cambridge, MA: Harvard University Press, 1964.
Luhmann, Niklas. *Law as a Social System*. Oxford: Oxford University Press, 2004.
Luskin, Robert C., et al. 'Considered Opinions: Deliberative Polling in Britain'. *British Journal of Political Science* 32, no. 3 (2002): 455–87.
MacDonald, Geoff, et al. 'Do People Use Reverse Psychology? An Exploration of Strategic Self-Anticonformity'. *Social Influence* 6, no. 1 (2011): 1–14.

Marmot, Michael. *Status Syndrome: How Your Social Standing Directly Affects Your Health and Life Expectancy*. London: Bloomsbury, 2004.
Martines, Lauro. 'Political Conflict in the Italian City States'. *Government and Opposition* 3, no. 1 (1968): 69–91.
Martines, Lauro (ed.). *Violence and Civil Disorder in Italian Cities 1200–1500*. Berkeley: University of California Press, 1972.
Mayr, Ernst. *The Growth of Biological Thought: Diversity, Evolution, and Inheritance*. Cambridge, MA: Belknap Press of Harvard University Press, 1982.
McManus, C. & Furham, A. 'Aesthetic Activities: Influences of Education, Background and Personality on Interest and Involvement in the Arts'. *British Journal of Psychology* 97, no. 4 (2006): 555–87.
McNeill, John R. *Something New under the Sun: An Environmental History of the Twentieth-Century World*. London: Penguin Books, 2000.
McNeill, John Robert, & McNeill, William. *The Human Web: A Bird's-Eye View of World History*. New York: W. W. Norton, 2003.
Melchionne, Kevin. 'Aesthetic Experience in Everyday Life: A Reply to Dowling'. *British Journal of Aesthetics* 51, no. 4 (2011): 437–42.
———, 'The Definition of Everyday Aesthetics'. *Contemporary Aesthetics*, 51, no. 4 (2011): 441.
Mennell, Stephen. 'Civilizing and Decivilizing Processes'. In *The Course of Human History: Economic Growth, Social Process, and Civilization*, edited by Johan Goudsblom et al., 101–16. New York: Routledge, 1996.
Milanovic, Branko. *Global Inequality: A New Approach for the Age of Globalization*. Cambridge, MA: Harvard University Press, 2016
———. *The Haves and Have-nots*. New York: Basic Books, 2016.
Milgram, Stanley. *Obedience to Authority: An Experimental View*. London: Pinter & Martin, 2010.
Mischel, Walter. *The Marshmallow Test: Mastering Self-Control*. New York: Little, Brown and Company, 2014.
Mischel, Walter, et al. 'Sustaining Delay Gratification over Time: A Hot-Cool System Perspective'. In *Time and Decision: Economic and Psychological Perspectives on Intertemporal Choice*, edited by George Lowenstein et al., 175–200. New York: Russell Sage Foundation, 2003.
Mithen, Steven. *The Singing Neanderthals. The Origin of Music*. London: Weidenfeld & Nicholson, 2005.
Molander, Per. *The Anatomy of Inequality: Its Social and Economic Origins and Solutions*. Brooklyn: Melville House Books, 2017.
———. *Condorcets misstag: hoten mot staten och demokratin*. Stockholm: Weyler, 2018.
Mollat, Michel, & Wolff, Philippe. *The Popular Revolutions of the Middle Ages*. London: George Allen & Unwin, 1973.
Moore, Barrington, Jr. *Injustice: The Social Bases of Obedience and Revolt*. London: MacMillan, 1978.
Morin, Edgar. *On Complexity*. Cresskill, NJ: Hampton Press, 2008.
Moscovici, Serge. 'Social Influence, Conformity Bias, and the Study of Active Minorities'. In *Advances in Experimental Psychology*, vol. 6, edited by Leonard Berkowitz, 149–202. New York: Academic Press, 1972.
Mousnier, Roland. *Peasant Uprisings in Seventeenth-Century France, Russia, and China*. New York: Harper & Row, 1970.

Nordforsk. *Adding Value to Nordic Research Cooperation: Recommendations for a New Initiative on Distribution of Health* . Oslo: Nordforsk, 2012.

Nyberg, Lars, et al., 'Memory Aging and Brain Maintenance'. *Trends in Cognitive Science* 16, no. 5 (2012): 292–305.

Nyhan, Brendan, & Reifler, Jason. 'When Corrections Fail: The Persistence of Political Misperceptions'. *Political Behavior* 32, no. 2 (2010): 303–30.

Nyhan, Brendan, et al. 'Effective Messages in Vaccine Promotion: A Randomized Trial'. *Pediatrics* 133, no. 4 (2015): e835–e842.

Nyhan, Brendan, et al. 'Taking Corrections Literally but Not Seriously? The Effects of Information on Factual Beliefs and Candidate Favorability'. *Political Behavior*, 42 (2020): 939–60. https://link.springer.com/article/10.1007/s11109-019-09528-x.

O'Donoughe, Ted, & Rabin, Matthew. 'Self-Awareness and Self-Control. In *Time and Decision: Economic and Psychological Perspectives on Intertemporal Choice*, edited by George Lowenstein et al., 217–44. New York: Russell Sage Foundation, 2003.

OECD. *Literacy in the Information Age*. Paris: OECD, 2000.

OECD. *Skills Outlook 2013*. Paris: OECD, 2013.

Oh, Hyung (David). 'Do Collectivists Conform More Than Individualists? Cross-Cultural Differences in Compliance and Internalization'. *Social Behavior and Personality* 41, no. 6 (2013): 981–94.

Oliner, S. P., & Oliner, P. M. *The Altruistic Personality – Rescuers of Jews in Nazi Europe: What Led Ordinary Men and Women to Risk Their Lives on Behalf of Others?* New York: Free Press, 1988.

Östling, Johan, et al. (eds). *Circulation of Knowledge: Explorations in the History of Knowledge*. Lund: Nordic Academic Press, 2018.

Östling, Johan, et al. 'The History of Knowledge and the Circulation of Knowledge: An Introduction'. In *Circulation of Knowledge: Explorations in the History of Knowledge*, edited by Johan Östling et al., 9–33. Lund: Nordic Academic Press, 2018.

Packer, Dominic J. 'On Being Both with Us and against Us: A Normative Conflict Model of Dissent in Social Groups'. *Personality and Social Psychology Review* 12, no.1 (2008): 50–71.

Pareliussen, Jan Kristian, et al. 'Income Inequality in the Nordics from an OECD Perspective'. In *Increasing Income Inequality in the Nordics. Nordic Economic Policy Review 2018*, edited by Rolf Aaberge et al., 17–57. Copenhagen: Nordic Council of Ministers, 2018.

Pennycook, Gordon, & Rand, David G. 'Who Falls for Fake News? The Roles of Bullshit Receptivity, Overclaiming, Familiarity, and Analytic Thinking'. *Journal of Personality*, 88, no. 2 (2019): 185–200. https://onlinelibrary.wiley.com/doi/full/10.1111/jopy.12476.

Pétrequin, Pierre. 'North Wind, South Wind: Neolithic Technical Choices in the Jura Mountains, 3700–2400 BC'. In *Technological Choices: Transformation in Material Cultures since the Neolithic*, edited by Pierre Lemonnier, 36–76. London: Routledge, 1993.

Piaget, Jean. *The Child's Conception of the World*. Totowa: Rowman & Littlefield, 1989.

———. *Six Psychological Studies* . New York: Vintage, Random House, 1968.

Piketty, Thomas. *Capital in the Twenty-First Century*. Cambridge, MA: Belknap Press of Harvard University Press, 2014.

Pinker, Steven. *The Better Angels of Our Nature: The Decline of Violence in history and Its Causes*. London: Penguin, 2011.

———. *How the Mind Works*. New York: W. W. Norton, 1997.

Pleijel, Hilding. *Hustavlans värld* . Stockholm: Verbum, 1970.

Plato. 'The Symposium'. In *The Portable Plato*, 121–90. Harmondsworth: Penguin Books, 1976.
Prelec, Drazen, & Bodner, Ronit. 'Self-Signaling and Self-Control'. In *Time and Decision: Economic and Psychological Perspectives on Intertemporal Choice*, edited by George Lowenstein et al., 277–300. New York: Russell Sage Foundation, 2003.
Quigley, Carroll. *The Evolution of Civilizations: An Introduction to Historical Analysis*. Indianapolis, Ind.: Liberty Fund, 1979.
Raj, Kapil. 'Networks of Knowledge, or Spaces of Circulation? The Birth of British Cartography in Colonial South Asia in Late Eighteenth Century'. *Global Intellectual History* 2, no. 1(2017): 49–66.
Rannells, Edward Warder. 'Aesthetic Expression and Learning'. *Journal of Aesthetics and Art Criticism* 5, no. 4 (1947): 314–20.
Reicher, Stephen, & Alexander, Haslam, S. 'On the Agency of Individuals and Groups: Lessons from the BBC Prison Study'. In *Individuality and the Group: Advances in Social Identity*, edited by Tom Postmes & Jolanda Jetten, 237–57. London: Sage, 2006.
Ricciardelli, Fabrizio. 'Violence and Repression in Late Medieval Italy'. In *The Culture of Violence in Renaissance Italy*, edited by Samuel Kline Cohn Jr & Fabrizio Riccardielli, 55–80. Florence: Le Lettere, 2012.
Richter, Gerhard. 'Walter Benjamin (1892–1940) and Theodor Adorno (1903–1969)'. In *Aesthetics: The Key Thinkers*, edited by Alessandro Giovannelli, 146–59. London: Continuum International, 2012.
Ridley, Matt. *Nature via Nurture: Genes, Experience & What Makes Us Human*. New York: HarperCollins, 2003.
Riksbankens Jubileumsfond. Åsa Wikforss. https://www.rj.se/en/anslag/2018/knowledge-resistance-causes-consequences-and-cures.
Ristin-Kaufman, Nuria, & Gullberg, Marianne. 'The Effects of First Exposure to an Unknown Language at Different Ages'. *Bulletin, Suisse de Linguistique Appliquée* 99 (2014): 17–29.
Roberts, Lissa. 'Situating Science in Global History'. *Itinerario* 33, no. 1 (2009): 9–30.
Rochat, François, & Modigliani, Andre. 'Captain Paul Grueninger: The Chief of Police Who Saved Jewish Refugees by Refusing to Do His Duty'. In *Obedience to Authority: Current Perspectives on the Milgram Paradigm*, edited by Thomas Blass, 91–110. Mahwah, NJ: Lawrence Erlbaum Associate, 2000.
Roine, Jesper. *Thomas Pikettys KAPITALET i det tjugoförsta århundradet*. Stockholm: Volante, 2014.
Saito, Yuriko. *Everyday Aesthetics*. Oxford: Oxford University Press, 2007.
Samuelson, Paul. 'Synthesis of the Principle of Acceleration and the Multiplier'. *Journal of Political Economy*, 47, no. 6 (1939): 786–97.
Sanders, Ronald. *The Days Grow Short: The Life and Music of Kurt Weill*. New York: Holt, Rinehart and Winston, 1980.
Schaffer, Simon. 'Newton on the Beach: The Information Order of *Principia Mathematica*'. *History of Science* 47, no. 3 (2009): 243–76.
Scheidel, Walter. *The Great Leveler: Violence and the History of Inequality from the Stone Age to the Twenty-First Century*. Princeton, NJ: Princeton University Press, 2017.
Schön, Lennart. *En modern svensk ekonomisk historia*. Stockholm: SNS, 2000.
Scruton, Roger. *Beauty: A Very Short Introduction*. Oxford: Oxford University Press, 2009.
Secord, James. 'Knowledge in Transit'. *Isis* 95, no. 4 (2004): 654–72.
Shachar, Ayelet. *The Birthright Lottery: Citizenship and Global Inequality*. Cambridge, MA: Harvard University Press, 2009.

Schellekens, Elisabeth. 'Immanuel Kant'. In *Aesthetics: The Key Thinkers*, edited by Alessandro Giovannelli, 61–74. London: Continuum International, 2012.
Sherif, Muzafer. *The Psychology of Social Norms*. New York: Harper & Bros, 1936.
Simon, Herbert O. 'The Architecture of Complexity'. In *Systems Thinking (vol. 1)*, edited by Gerald Midgley. London: Sage, 2003, 306–408.
Silverbark, Thord. *Fysikens filosofi: Diskussioner om Einstein, relativitetsteorin och kvantfysiken i Sverige 1910–1970*. Stockholm: Brutus Östlings förlag, 1999.
Snow, C. P. *The Two Cultures*. London: Cambridge University Press, 1993.
Sobel, Dava. *Galileo's Daughter: A Drama of Science, Faith, and Love*. London: Fourth Estate, 1999.
Sorokin, Pitirim. *Social and Cultural Dynamics: A Study of Change in Major Systems of Art, Truth, Ethics, Law and Social Relationships*. Boston, MA: Porter Sargent, 1957.
Soroush, Vosoughi, et al. 'The Spread of True and False News Online'. *Science* 359, no. 6380 (2018): 1146–51.
Stavrianos, Leften S. *Lifelines from Our Past: A New World History*. Armonk: M. E. Sharpe, 1997.
Stecker, Robert. 'Plato'. In *Aesthetics: The Key Thinkers*, edited by Alessandro Giovannelli, 8–20. London: Continuum International, 2012.
Steinberg, Laurence, & Monahan, Kathryn C. 'Age Differences in Resistance to Peer Influence'. *Developmental Psychology* 43, no. 6 (2007): 1531–43.
Strimling, Pontus, et al. 'The Connection between Moral Positions and Moral Arguments Drives Opinion Change'. *Nature Human Behaviour* 3, 15 July (2019): 922–30.
Stringhini, Silvia, et al. 'Socioeconomic Status and the 25 x 25 Risk Factors as Determinants of Premature Mortality'. *Lancet*, no. 389 (2017): 1229–37. http://www.thelancet.com/pdfs/journals/lancet/PIIS0140-6736(16)32380-7.pdf.
Subramanyam, Malavika, et al. 'Relative Deprivation in Income and Self-Rated Health in the United States'. *Social Science & Medicine* 69, no. 3 (2009): 327–34.
Swire, Briony, et al. 'Processing Political Misinformation: Comprehending the Trump Phenomenon'. *Royal Society Open Science* (1 March 2017). https://royalsocietypublishing.org/doi/full/10.1098/rsos.160802.
Taber, Charles S., & Lodge, Milton. 'Motivated Skepticism in the Evaluation of Political Beliefs'. *American Journal of Political Science* 50, no. 3 (2006): 755–69.
Tainter, Joseph. *The Collapse of Complex Societies*. Cambridge: Cambridge University Press, 1988.
Tawney, R. H. *Equality*. London: George Allen & Unwin, 1964.
Taylor, Charles. *The Ethics of Authenticity*. Cambridge, MA: Harvard University Press, 1991.
Tice, Dianne M. 'Self-Concept Change and Self Presentation: The Looking Glass Self Is also a Magnifying Glass'. *Journal of Personality and Social Psychology* 63, no. 3 (1992): 435–51.
Tomasello, Michael. *The Cultural Origins of Human Cognition*. Cambridge, MA: Harvard University Press, 1999.
Tourraine, Alain. *The Postindustrial Society. Tomorrow's Social History: Classes, Conflicts and Culture in the Programmed Society*. New York: Random House, 1971.
Toynbee, Arnold J. *A Study of History: Abridgement of Volumes I-VI by D.C. Somervell*. Oxford: Oxford University Press, 1974.
Triandis, Harry C. 'Individualism–Collectivism and Personality'. *Journal of Personality* 69, no. 6 (2001): 907–24.
———. 'The Self and Social Behavior in Differing Cultural Contexts'. *Psychological Review* 96, no. 3 (1989): 506–20.
Turchin, Peter. *War and Peace and War: The Rise and Fall of Empires*. London: Penguin Books, 2006.

Turchin, Peter, & Nefedov, Sergey A. *Secular Cycles*. Princeton, NJ: Princeton University Press, 2009.

Turchin, Peter, et al. 'Quantitative Historical Analysis Uncovers a Single Dimension of Complexity That Structures Global Variation in Human Social Organization'. *PNAS* (9 January 2018). https://www.pnas.org/content/115/2/E144.short.

Urrutia, Doris Elin. 'Brazil's Space Agency Head Was Forced Out for Defending Climate Science'. 18 June 2018. https://www.space.com/brazil-space-agency-leader-dismissed.html.

van der Leeuw, Sander. 'Giving the Potter a Choice'. In *Technological Choices: Transformation in Material Cultures since the Neolithic*, edited by Pierre Lemonnier, 238–88. London: Routledge, 1993.

V-Dem. 'The Third Wave of Autocratization Is Here'. 11 March 2019. https://www.v-dem.net/en/news/third-wave-autocratization-here/.

Watts, Tyler W., et al. 'Revisiting the Marshmallow Test: A Conceptual Replication Investigating Links between Early Delay of Gratification and Later Outcome'. *Psychological Science* 29, no. 7 (2018): 1159–77.

Westbrook, Michael H. *The Electric Car: Development and Future of Battery, Hybrid and Fuel-Cell Cars*. London: Institution of Electric Engineers, 2001.

Whitehead, Alfred. *Science and the Modern World*. New York: Macmillan, 1925.

Wikipedia. 'Igor Stravinskij'. https://sv.wikipedia.org/wiki/Igor_Stravinskij.

Wikipedia. 'Taipeil 101'. https://en.wikipedia.org/wiki/Taipei_101.

Wittrock, Björn. 'The Meaning of Axial Age'. In *Axial Civilizations and World History*, edited by Johann P. Arnasson et al., 51–86. Leiden: Brill, 2005.

Wootton, David. *The Invention of Science: A New History of the Scientific Revolution*. New York: HarperCollins, 2015.

Zimbardo, Philip G. 'On Rethinking the Psychology of Tyranny: The BBC Prison Study'. *British Journal of Social Psychology* 45, no. 1 (2006): 1–40.

Zunshine, Lisa. *Why We Read Fiction: Theory of the Mind and the Novel*. Columbus: Ohio State University Press, 2006.

INDEX

academic community 92, 97
aesthetic
  action 84, 86
  activities 83, 87, 88, 104, 105
  appearance 83
  appreciation 80
  attractiveness 81
  concerns 83, 87
  considerations 79, 83–84, 86, 87
  design 83
  endeavours 85–86, 88, 105
  experience 84, 86–87, 93, 101, 104, 127
  intentions 79, 83, 86, 105
  investment 84
  need 80, 84
  qualities 80–81
  scholars 85–86
aesthetics 81–82, 84–88, 105
affluent worker 64
Alfani, Guido 73
altruism 121n35
animals 44, 81, 105
anthropological research 50
anthropologists 62, 70, 97
anti-egalitarian 67, 75
Antiquity 75
anti-scientific attitude 93
archaeologists ix, x, 83
archaeology 96, 130
Arendt, Hanna 49–50
art historians 85, 105
artistic research 93–94
Asch, Solomon 47–48, 50, 58
astronomers 97
authoritarian rule 10
autocratisation 20

autonomous individuals 46, 51, 106
Axial Age 76, 107, 108

beauty 55, 80, 82–85
behavioural
  attitudes 65
  patterns 57
bell curve 71
Bell, Daniel 18
biologists 93, 97
  green 97
  human 52
Boehm, Christopher 70–71, 75
Bourguignon, François 72
brain imaging 96
Braudel, Fernand 36
Browning, Christopher 49–50
bureaucratic apparatuses 70
business cycles 35

capital–income ratio 71
central administration 44, 108
centralised state 62
chemists 91–92
circular change 35, 109
circulation perspective 23
citizens 6, 18, 20–21, 27, 43–45, 56–58, 62, 71, 77, 92–93, 104, 108, 111
civilising process 62
class consciousness 64
climate 36, 97
  change 58, 94
  crisis 45
  system 95
climatologists 95
cognitive
  change 106

cognitive (*cont.*)
  development 10
  factors 103
  sciences 97
collapse 56–57, 89
collectivism 37, 45, 57, 108
comparative advantages 91
  approach 13, 26, 63
  findings 50
  research 45, 106
  studies 110
complexity 21, 26, 29, 30, 34, 46, 92, 95, 109, 112
Comte, August 18
confirmation bias 25
conflicting interests 46
conformity research 25
consumption patterns 64
Copernicus, Nicolaus 24
creative minority 52
creativity 32, 101, 111
critics of society 92–93
cultural
  change 29, 30, 34–35, 47, 97, 106, 109
  clashes 31
  climate 36
  conditions 64
  convergence 31
  de-centrism 60
  development 52
  differences 49, 95
  dynamics 46, 52
  evolution 19, 30, 52, 89, 90
  expressions 88
  processes 45, 53
  roadblocks 40
  selection 101
  system 30
  traits 36
  variance 48
  variation 53
culture(s) ix, 2, 7, 17, 30, 50–51, 55, 84, 95, 97, 101, 107
  collectivist 48
  consumption of 35, 57, 94
  convergence of 30
  high 85
  human 62, 83

individualist 48
  of narcissism 58
  of science 6
  of violence 62
  performative 92
  research of 4, 94
cumulative 31, 34, 41, 109
  change 34, 36–37, 109
  development 40, 52
  growth/progress 4, 10
  novelties 36
  process 32, 35
cumulativeness 31, 40
curiosity 11–12, 92, 94
cyclical processes 35, 109

Dark Ages 63
Darwin, Charles 23–24, 96
Darwinian-Mendelian synthesis 8
de-centrism 60
deliberate design 83
Delumeau, Jean 63, 65
democracy 10
democratic
  countries 20
  empowerment 20
  rule 10
  society 50
demographic collapses 57
desire(s) 17, 44–45, 47, 55–56, 61–62, 65, 78, 85, 89, 106–7, 110
determinism 40
Dewey, John 81–82, 85–86, 87
Di Tullio, Matteo 73
disciplining of manners 59
disconfirmation bias 25
discount 61
  functions 59
  rates 59, 60
discoveries 1, 22, 24, 95
dissolution of the state 73
distribution
  of income 35, 56, 68, 71–72
  of knowledge 19
  of pleasure 55–56
  of resources 69–70, 72, 77, 111, 112
  of wealth 73, 68
  unequal 68, 111–12
DNA analysis 90

Dorling, Daniel 58, 69, 74
Drucker, Peter 18
Duby, Georges 75
Duerr, Hans Peter 62
Dworkin, Ronald 77
dynamism of human society 30–31

economic
 activity 35
 circumstances 56
 conditions 64
 development 35, 72, 109
 egalitarianism 77
 gaps 57
 growth 57, 71–72
 historian(s) 56, 70
 history 71
 problems 45, 108
 research 65
 spheres 60
 troubles 60
 theory 59
 waves 36
economist(s) x, 36, 59–60, 70–71
educational process 62
egalitarian 67–68, 70, 74–77
egalitarianism 67–68, 70, 75, 76–77
egocentrism 60
Eichmann, Adolf 49–50
Eisner, Manuel 62
Ekman, Marie-Louise 81
Elias, Norbert 62–63, 65
elites 23, 56, 70
emotional harmony 85
empathy 56
empirical
 correctives 87
 inquiries 85
 investigations 85
 operationalisation 85
 research 86–87, 96
 scrutiny 87
 studies 87, 111
energy regime 57
enlightenment 75
environmental humanities 90
epidemics 73
epigenetics 8

epistemology 96
equality x, 67–70, 72–73, 77
 archaic 70
 before the law 74, 76, 112
 social 74
appreciation 80–81, 83
ethnic groups 50
everyday 86
 aesthetics 80, 84, 86–88, 105, 111
 beauty 85
 expressions 107
 life/lives 80, 86
 utensils 81
evolution ix, 19, 23, 30, 32, 35, 48, 52, 76, 89–90, 96, 106, 112
evolutionary
 biology 8
 curve 70
 idea 87
 oriented 62
 process 45
 research 50
 significance 53, 85
 terms 87
 theory 96
existential 31, 34, 88
experimental
 findings 60
 interventions 65
 music 34
 studies 26, 50, 97
experiment(s) 10, 25, 48–49, 53, 60–61, 110
experts 20
extra-scientific human need 92

facts 10, 74, 94–95, 105
fallibilist attitude 95
fashion 3, 19, 22, 35
fine arts 80, 86–88, 105
Flynn effect 19
Fogel, Robert 56, 64
free will 92–93
freedom
 of choice 40
 of expression 20, 76
 of the press 20
Freud, Sigmund 60, 79
Fry, Roger 84

Galileo, Galilei 23–24
gender
  gap 68
  research 90
general truths 5
generalisations 5, 9, 10, 48, 50
genetics 8
Geyer, Anne 48–49
Gini
  Coefficient 72
  model 69
  points 71
Golden Rule 56, 76, 108
Goldthorpe, John 64
grand narratives 4
gratification 58, 61, 64, 110
great chain of being 74
group
  identities 79
  pressure x, 47, 49–52
guilt 63

Haslam, Alexander 49
Hayek, Friedrich 77
heliocentrism 23, 24
heterarchy 49
hierarchy(ies) 70
Higgs boson 95
Hilbert, David 7, 13, 90
historians x, 13, 23, 26, 55, 57, 59, 62–63, 65, 70, 85, 96–97, 105
historical
  analyses 53
  change 36, 109
  context 17
  destiny 63
  fluctuations 91
  investigations 26
  periods 84
  perspective 22, 65
  presentations 49
  process 44, 57
  questions 61
  research 45, 61, 65, 106
  studies 27, 50, 65, 110
  trajectory 63
  variance/variation 48, 50, 53

history
  of humankind 57
  of mentalities 63
  of science 23–24, 95
  of technical growth 111
  of the world 71
Hodges, Bert 48–49
holistic approach 65
Homo sapiens 30
homogenisation 30
human
  action 35, 43, 104, 109
  behaviour 64, 95
  brain 19, 97
  civilisation(s) 57, 65
  condition ix, 2, 4, 7, 9, 11, 91, 92, 94, 96, 111
  genome 19, 29, 109
  interaction 29, 31, 43, 45, 104, 108–9
  mind 83, 87
  misbehaviour 63
  needs 8, 36, 44–45, 62, 65, 80, 84, 89, 94, 106, 109
  prehistory 83
  society 30, 31, 35–36, 46, 52, 67, 90, 106, 109, 112
humanist values 92
humanities research 3–4, 90
humankind 11, 17, 31, 57, 71, 75–77, 90, 91, 94, 108–9, 111

ideology(ies) 35, 75, 109
immaterial resources 56
impulses 56, 61–62
incest 63
income
  distribution 35, 56, 71–72
  inequality 68, 72–74
  gaps 56, 71–74
individualism 37, 45, 57, 108
industrial
  capitalism 57, 71, 77
  period 71
  revolution 57
  sector 20
  societies 71
  workers 57

inegalitarianism 76
inequality 56, 58, 67–77, 112
information 18, 19, 25–26, 44, 72, 104
informational conformity 48
innovation(s) 22, 31–32, 34, 36, 40, 52, 109, 111
institutional
   change 40
   constructs 31
   preconditions 111
   reforms 36
   rejuvenation 52
   sense 80
intellectual
   history 25, 85
   investment 58
intelligent design 35, 109
interactionism 49
interdisciplinary approach 91
interpretations 1, 48, 92, 94, 95
intertemporal choice 59

Kandel, Eric 94
Kant, Immanuel 80
Keynes, John Maynard 35
Keynesian 36, 56
knowledge
   advancement 23, 78, 89, 112
   affirming attitude 21–22
   breakthroughs 27
   cartographic 23
   counter-intuitive 13, 103
   destiny of 26
   dissemination of 20
   distribution of 19
   dynamics of 23
   economy 18
   Eurocentric perspective on 23
   historians 23
   mass of 20
   nodes of 23
   production of 6, 23
   progress 3, 13
   rejection of 17, 103
   resistance to 17, 89
   societal significance of 89
   society 18, 19, 20–22, 26
   stock of 9, 18–20, 104
   traits of 17–18, 103
   true 20–21, 25–26
Kondratieff, Nikolai 36
Kuznets, Simon 71–72

Lane, Robert 18
large-scale communities 48
Lasch, Christopher 58
laws of nature 81
Le Goff, Jacques 63, 65
Le Pen, Marine 25
Leddy, Thomas 84
Lee, Jessica 80
let-go system 61
linguistic topics 90
linguist(s) x, 10
Lister, Joseph 24
literary
   scholars 85, 92
   studies 96
living conditions 57, 64, 74
long waves 35
longitudinal studies 10
long-term approach 65
López-Claros, Augusto 68
Lord, Charles 25

macro processes 31
macro-historical
   analyses 53
   approach 65
   perspective 22
majority(ies) 3, 20, 34, 46, 48–51, 53, 106, 122n39
   conformist 46, 106
   contemporary 23, 51
   global 47
   local 48, 53
   past 51
Malthus, Thomas Robert 57
Malthusian checks 57
marshmallow experiments 60, 110
Marxist 77
mass consumption 56, 63
material
   circumstances 63

material (*cont.*)
  conditions 110
  resources 68, 72, 111
mathematical questions 7
mathematician(s) 7, 18, 22, 46,
    80, 90
mathematics 7, 22, 32
Matthew effect 68
media studies 27
medical care 68
Melchionne, Kevine 86
memory
  episodic 10
  semantic 10
mercantilism 75
methods 13, 96, 106
  qualitative 92
  quantitative 92
  statistical 95
Middle Ages 22, 63, 75
migration flows 90–91
Milanovic, Branko 71, 72
Milgram, Stanley 47, 49, 50
mind-reading 85
minority(ies) 2–3, 22–23, 26, 46–47,
    50–53, 64, 77, 106
Mischel, Walter 60, 61, 65, 110
misinformation 20
Mithen, Steven 85
Molander, Per x, 68–69, 71, 73
Morrion, Christian 72
musical
  change 34
  compositions 32, 34
  development 34
  traditions 34
  trends 34
musicologists 85, 105
Myrdal, Janken ix–x, 11–12

Nakhjavani, Bahiyyih 68
narcissism 79
narrative(s) 4, 9
natural
  constraints 40
  laws 93
  sciences 3, 13, 94, 96
  scientists 90, 93, 95
  selection 96

networks 31, 109
neurosciences 97
neurology 96
Newton, Isaac 23–24
nobility 62
non-aesthetic function 80
non-aesthetic needs 80
non-art objects 80
non-circular change 35
non-conformity research 25
non-cumulative change 35
non-expert 20
non-physical representations 82
non-state societies 70
non-technical considerations 40
norm
  breakers 44
  change 44–46, 89
normative
  conformity 48
  shifts 44

observational studies 26
obsolete technology 40
opportunism 48, 51
opportunity cost 55
over-consumption 58

palaeontologists 97
Pareto, Vilfredo 77
Particularism 96
path dependency 40
philosophers 77, 85, 96
philosophy 11, 27
phoneticians 96
physical
  conditions 40
  power 43, 104
  representations 82
  suffering 57
  transportation 37
  violence 46, 112
physicists 9, 91
physics 1, 8, 40, 97
Piaget, Jean 60
Piketty, Thomas 71, 72, 73
Pinker, Steven 62, 85
plasticity 19, 97
Pogge, Thomas 77

policymakers 7, 93
postmodernists 4
political science 27
pollution 57
Popperians 21
popular revolts
post-knowledge society 19
postponement 55, 59, 63, 64, 110
post-war years 74
preference(s) 45, 59, 63, 84, 89
prehistory 65, 75, 83
pre-industrial 62, 71
primary sector 64
primates 70
private consumption 57
professional art 79
professional dancer 83, 105
psychodynamic theory 60
psychological research 48, 49, 110
psychologist(s) x, 10, 13, 25, 49, 55, 60, 62, 85, 96
psychology 27, 47–48, 60, 97

quantum mechanics 8, 9

Rawls, John 77
redistributive mechanisms 69
reductionism 13
regulated interaction 56
Reicher, Stephen 49
relativist attitude 95
reliability 96
reverse dominance hierarchy 70
revolutionary upheavals 73
Rift Valley 30
rights
  human 44, 76–77, 108
  national 77
risk aversion 61, 110
Roemer, John 72
rules
  of logic 96
  of science 94

Saito, Yukiko 83
Samuelson, Paul 36
Scheidel, Walter 72–73
Schneider, Harold 70
scientific
  community 90, 93
  curiosity 92
  expertise 92
  fields 95
  preferences 89
  idiocyncracies 89
  intellectuals 94
  playground 91
  progress 9
  scrutiny 11
  students 107
  studies 93
Scruton, Roger 55, 80, 85, 87
Second World War 9
sedentary life 31, 34
self-
  constraint 65
  control 59, 60–65, 110
  restraint 60–61, 63
semantics 96
sequential thinking 29, 44, 104
sexual selection 87
Shachar, Ayelet 77
Sherif, Muzafer 47
Singer, Peter 77
small-scale communities 48
sociological studies 64
sociologist(s) 18, 55, 62, 64
specialisation 18, 34, 104, 111
stable societal conditions 73
standing armies 70
state-based civilisations 70
static society 57
Stravinsky, Igor 34
subsistence 75
  economy 34
  level 64, 68
symbolic representations 81
systematic experiments 13

Tawney, Richard Henry 67, 72
technical/technological
  avenue 36
  change 37, 40, 111
  choices 40
  development 20, 32, 40, 111
  growth 111
  inertia 40
  path dependency 40
  progress 89

technical/technological (*cont.*)
   rejuvenation 52
   revolution 72
   sciences 97
   steps 40
textual analysis 96
Tourraine, Alain 18
Toynbee, Arnold 52
transcendent thoughts 76, 108
transmission 52
   horizontal 43
   vertical 43
tribal societies 70
trickle-down perspective 23
Trump, Donald 25, 26
truth 1–2, 17, 24
truth-affirming attitude 3
Turkey 20

unity of sciences 13
universal personhood 77
universalistic perspective 77
utilitarian objects 80

validity 76, 96, 108
values 25, 46, 49, 52, 63, 84, 92

variation 5, 26, 30, 35, 50, 53, 61, 68, 95, 96
violence 44–46, 62, 73, 89, 104–5, 112
visual impressions 83

Wallace, Alfred Russel 96
wartime 71
Weill, Curt 34
welfare 77, 110
   institutions 44, 108
   project 77
   society 57, 64
   system 57
Western world 18, 23
Whitehead, Alfred 18
Workforce 20–21
working
   class 64
   life 21, 64
   people 76
   society 75
world historian 13
worldwide justice 77
writing systems 32

Zimbardo, Philip 47, 49, 50
Zunshine, Lisa 85

www.ingramcontent.com/pod-product-compliance
Lightning Source LLC
Chambersburg PA
CBHW061451300426
44114CB00014B/1932